Extracellular Matrix - Developments and Therapeutics

Edited by Rama Sashank Madhurapantula, Joseph Orgel P.R.O. and Zvi Loewy

Published in London, United Kingdom

IntechOpen

Supporting open minds since 2005

Extracellular Matrix - Developments and Therapeutics
http://dx.doi.org/10.5772/intechopen.77848
Edited by Rama Sashank Madhurapantula, Joseph Orgel P.R.O. and Zvi Loewy

Part of IntechOpen Book Series: Biochemistry, Volume 23
Book Series Editor: Miroslav Blumenberg

Contributors
Maryada Sharma, Kavita Kaushal, Sanjay Singh Rawat, Manjul Muraleedharan, Seema Chhabra, Nipun Verma, Anupam Mittal, Ajay Bahl, Madhu Khullar, Anuarg Ramavat, Naresh Panda, Fernando Rodriguez-Pascual, Nilkamal Pramanik, Sameer Kumar Jagirdar, Jason P. Hodde, Michael C. Hiles, Robert Weiss, Alexander Daoud, Aman Karim, Muhammad Zeeshan Bhatti, Elena Vecino, Xandra Pereiro, Noelia Ruzafa, Ane Zulueta, Alfredo Sarmiento, Alejandro Diez, Zhijie Wang, Kristen LeBar

First published in London, United Kingdom, 2021 by IntechOpen
IntechOpen is the global imprint of INTECHOPEN LIMITED, registered in England and Wales, registration number: 11086078, 5 Princes Gate Court, London, SW7 2QJ, United Kingdom
Printed in Croatia

British Library Cataloguing-in-Publication Data
A catalogue record for this book is available from the British Library

Additional hard and PDF copies can be obtained from orders@intechopen.com

Extracellular Matrix - Developments and Therapeutics
Edited by Rama Sashank Madhurapantula, Joseph Orgel P.R.O. and Zvi Loewy
p. cm.
Print ISBN 978-1-83968-235-3
Online ISBN 978-1-83968-236-0
eBook (PDF) ISBN 978-1-83968-237-7
ISSN 2632-0983

For EU product safety concerns:
IN TECH d.o.o., Prolaz Marije Krucifikse Kozulić 3, 51000 Rijeka, Croatia, info@intechopen.com or visit our website at intechopen.com.

IntechOpen Book Series
Biochemistry
Volume 23

Rama Sashank Madhurapantula is a research assistant professor in the Biology Department, Illinois Institute of Technology, Chicago. His current research involves developing microscopy techniques to establish macroscopic stress vs strain relations in body tissues that present mixed-tissue compositions, in conjunction with X-ray diffraction scanning techniques to establish tissue composition. His doctoral work was in understanding molecular changes to collagens with diseases and changes in fibrous structures such as myelin. In recognition of scientific endeavors and achievements, Dr. Madhurapantula was elected as chair to the Fiber Diffraction Special Interest Group of the American Crystallographic Association in 2020 and awarded the Margaret C. Etter Student Lecturer award in 2014.

Professor Joseph Orgel is a British American scientist based at the Illinois Institute of Technology (Illinois Tech), Chicago, with appointments in biology and biomedical engineering. His research explores fundamental structural biochemistry underlying disease and possible treatments. Using novel techniques, Dr. Orgel and his group have been able to visualize the molecular organization of connective and neurological tissues at nanometer (or better) resolution. He leads investigations of diseases such as Alzheimer's, traumatic brain injury, heart disease, and arthritis in collaboration with the US Army. An awardee of the NSF CAREER award, he has been Biochemistry Section Editor of PLOS ONE since 2008. In early 2021, he was named Vice Provost for Academic Affairs at Illinois Tech.

Dr. Zvi Loewy is a senior academic leader and an experienced global pharmaceutical–biotechnology executive. He leverages a diversified background in big-pharma senior management, biotech startup creation, and academia. Dr. Loewy has served as a board member of the New Jersey Bioscience Center Incubator since 2010. From 2005 to 2020 he was a board member of the Jerusalem College of Technology. His international experience has included leading international research teams; championing the penetration and commercial launch of healthcare products worldwide; and leading open innovation in the Mideast. Dr. Loewy received his BA from Yeshiva University, New York, his MS from Rensselaer Polytechnic Institute, New York, and his Ph.D. in Molecular Biology from the Albert Einstein College of Medicine, New York. He has more than twenty-five issued patents to his credit.

Editors of Volume 23:
Rama Sashank Madhurapantula
Illinois Institute of Technology, Chicago, USA

Joseph Orgel P.R.O.
Illinois Institute of Technology, Chicago, USA

Zvi Loewy
New York Medical College, USA

Book Series Editor: Miroslav Blumenberg
NYU Langone Medical Center, New York, USA

Scope of the Series

Biochemistry, the study of chemical transformations occurring within living organisms, impacts all of the life sciences, from molecular crystallography and genetics, to ecology, medicine and population biology. Biochemistry studies macromolecules - proteins, nucleic acids, carbohydrates and lipids –their building blocks, structures, functions and interactions. Much of biochemistry is devoted to enzymes, proteins that catalyze chemical reactions, enzyme structures, mechanisms of action and their roles within cells. Biochemistry also studies small signaling molecules, coenzymes, inhibitors, vitamins and hormones, which play roles in the life process. Biochemical experimentation, besides coopting the methods of classical chemistry, e.g., chromatography, adopted new techniques, e.g., X-ray diffraction, electron microscopy, NMR, radioisotopes, and developed sophisticated microbial genetic tools, e.g., auxotroph mutants and their revertants, fermentation, etc. More recently, biochemistry embraced the 'big data' omics systems.

Initial biochemical studies have been exclusively analytic: dissecting, purifying and examining individual components of a biological system; in exemplary words of Efraim Racker, (1913 –1991) "Don't waste clean thinking on dirty enzymes." Today, however, biochemistry is becoming more agglomerative and comprehensive, setting out to integrate and describe fully a particular biological system. The 'big data' metabolomics can define the complement of small molecules, e.g., in a soil or biofilm sample; proteomics can distinguish all the proteins comprising e.g., serum; metagenomics can identify all the genes in a complex environment e.g., the bovine rumen. This Biochemistry Series will address both the current research on biomolecules, and the emerging trends with great promise.

Contents

Preface

The elusiveness and intrigue of human physiology and the appeal to understand it better is a major source of inspiration for millions of scientists and physicians worldwide. To understand a machine as complex as the human body one must look at the various elements that make up the machine. Historically, major emphasis has been placed on understanding cellular processes and their contribution to physiology and pathology. Over the last thirty-five years, the advent of and easier access to new imaging techniques and methods to understand molecular structure and function (various microscopy methods, X-ray diffraction and scanning, X-ray fluorescence, etc.) have made it much easier to understand sub- and intra-cellular processes, leading to the need for us to understand the extracellular matrix (ECM).

After initially considering the ECM as a simple connective tissue that literally connects things in the body, research has brought us to understand the complexity and the intelligence of the various ECM constituents. Understanding the molecular and packing structures of various fibrillar collagens (the most prevalent protein in mammals) has been made possible through synchrotron X-ray diffraction experiments. Changes to native collagen structures because of disease as well as a mechanical and chemical insult are also now more readily explored and explained as a result of these techniques. They also bring to the fore the importance of the physical, chemical, and biological attributes that trigger, restrict, moderate, and modify intracellular pathways in both good and bad ways. For instance, the time-sensitive activation of discoidin domain receptors by fibrillar collagens is directly linked with various physiological and pathological intracellular cascades [1].

Other ECM elements associated with fibrillar collagen have also become major targets for various therapeutic and surgical applications. Hyaluronic acid (HA), one of the chief elements of the ECM, has garnered significant attention and research into its biology and therapeutic applications. Several advancements and contributions have been made in the use of HA in wound dressings, bone and cartilage regeneration, cosmetics, and food applications in just the last five years.

We are at a time when understanding biology and applying it to biomedical interventions is becoming faster and we are generating more interesting and promising applications every day. I was highly enthused when IntechOpen reached out to me about serving as the editor for this book. I have spent the last eleven years of my life, through doctoral and postdoctoral studies, in the area of ECM biology, and I am more intrigued by the complexity and challenges it presents. This book is a collection of some of the more recent developments in the areas of collagen and hyaluronic acid structure and function. I would like to thank all the authors and their teams for working on this project with me and submitting their amazing manuscripts. It was genuinely rewarding to have reviewed and edited your work.

I would also like to thank my co-editors, Prof. Joseph Orgel from the Illinois Institute of Technology and Prof. Zvi Loewy from the Touro College of Pharmacy and New York Medical College. Special thanks to Prof. Orgel for being my mentor and helping me become a scientist.

I would like to express my humble gratitude to the staff at IntechOpen for giving me this opportunity to work on this exciting project. Special thanks to Author Service Manager Ms. Marijana Francetic and commissioning editor Ms. Jelena Germuth for their constant support, encouragement, patience, and perseverance.

I cannot describe in words how grateful I am to my wife Mounika for her constant support and encouragement through all my endeavors. I would like to dedicate this book to my parents, Venkata Ramana Murthy and Rajya Lakshmi Madhurapantula. You instilled in me the values of hard work and a strong education. Your love and support have given me the motivation to choose this field and become a better human being.

Rama Sashank Madhurapantula
Research Assistant Professor,
Department of Biology,
Illinois Institute of Technology,
Chicago, USA

Joseph Orgel P.R.O.
Illinois Institute of Technology,
United States of America

Zvi Loewy
New York Medical College,
USA

References

[1] Orgel JP, Madhurapantula RS. A structural prospective for collagen receptors such as DDR and their binding of the collagen fibril. Biochimica et Biophysica Acta (BBA)-Molecular Cell Research. 2019 Nov 1;1866(11):118478.

Section 1

Extracellular Matrix - Structure and Function in Physiology, Pathology and Medicine

Chapter 1

The Cellular Stress Response Interactome and Extracellular Matrix Cross-Talk during Fibrosis: A Stressed Extra-Matrix Affair

Maryada Sharma, Kavita Kaushal, Sanjay Singh Rawat, Manjul Muraleedharan, Seema Chhabra, Nipun Verma, Anupam Mittal, Ajay Bahl, Madhu Khullar, Anurag Ramavat and Naresh K. Panda

Abstract

Diverse internal and external pathologic stimuli can trigger cellular stress response pathways (CSRPs) that are usually counteracted by intrinsic homeostatic machinery, which responds to stress by initiating complex signaling mechanisms to eliminate either the stressor or the damaged cells. There is growing evidence that CSRPs can have context-dependent homeostatic or pathologic functions that may result in tissue fibrosis under persistence of stress. CSRPs can drive intercellular communications through exosomes (trafficking and secretory pathway determinants) secreted in response to stress-induced proteostasis rebalancing. The injured tissue environment upon sensing the stress turns on a precisely orchestrated network of immune responses by regulating cytokine-chemokine production, recruitment of immune cells, and modulating fibrogenic niche and extracellular matrix (ECM) cross-talk during fibrotic pathologies like cardiac fibrosis, liver fibrosis, laryngotracheal stenosis, systemic scleroderma, interstitial lung disease and inflammatory bowel disease. Immunostimulatory RNAs (like double stranded RNAs) generated through deregulated RNA processing pathways along with RNA binding proteins (RBPs) of RNA helicase (RNA sensors) family are emerging as important components of immune response pathways during sterile inflammation. The paradigm-shift in RNA metabolism associated interactome has begun to offer new therapeutic windows by unravelling the novel RBPs and splicing factors in context of developmental and fibrotic pathways. We would like to review emerging regulatory nodes and their interaction with CSRPs, and tissue remodeling with major focus on cardiac fibrosis, and inflammatory responses underlying upper airway fibrosis.

Keywords: extracellular matrix, homeostasis, tissue repair, fibrosis, cellular stress, RNA binding proteins, RNA interactome

1. Introduction

Fibrosis is an inherent reparative response invoked to restore tissue integrity following a pathologic insult, which metamorphoses into a devastating pathology culminating in scars due to self-catenating heralding inflammatory loops. Regeneration is a fundamental biological process initiated to orderly replace the damaged tissues, however, deregulation of chronic inflammation, growth factor-receptor cross-talk, intra-intercellular communication, and various extracellular matrix proteins eventually result in an aberrant wound healing response marked by fibrosis or scarring. Excessive scarring can obliterate tissue architecture, culminating in organ failure and death. Therefore, lack of coordination and synchronization in molecular and cellular events that guide *"regeneration"* results in *"degeneration"* of the affected organ. Fibroproliferative disorders are widely occurring and include pulmonary fibrosis, systemic sclerosis, liver cirrhosis, cardiovascular disease, progressive kidney disease, corneal scarring, proliferative vitreoretinopathy and posterior capsular opacification. Aberrant tissue remodeling marked by fibrosis is also implicated in cancer metastasis and chronic graft rejection in transplant recipients. The obvious deprecating impacts of fibrosis are enormous deterrents to patients; moreover, the disease has failed to meet the required treatments till date. Lack of availability of desired therapeutic interventions is majorly due to an incomplete understanding of the mechanism of the disease, therefore, gaining insights into the mechanistic pathways of fibrosis would facilitate improved therapeutic approaches to target novel mediators besides the cryptic or altered ECM components as previously reported by our group [1–5].

TGF-β is a central node in driving fibrotic pathways and other diseases, however, targeting TGF-β pathways has not met desirable clinical success, perhaps due to the incomplete mechanistic information on its role in development and pathology. Therefore, gaining insights into regulation of TGF-β during normal development and pathology could facilitate recognition of alternate target-pathways that may spare or minimally perturb the role of TGF-β in physiology. An interesting and relatively less explored theme is involvement and regulation of TGF-β isoforms in fetal wound healing that is marked by absence of scar. Intriguingly, there are common mediators and unifying pathways that underlie tissue repair, homeostasis and fibrosis in diverse organs; therefore, harnessing the potential non-fibrotic themes from scarless wound healing might add to the understanding of challenging fibrotic disorders. A systematic and meticulous re-assessment and re-evaluation of the role of mediators of scarlessly healing wounds might offer a reasonably potential tool to be manipulated to prevent fibrosis and decode the invisible lines dividing the "homeostasis-tissue repair-fibrosis continuum".

Interestingly, there are few physiological paradigms where wounds heal scarlessly or with minimal scarring, for instance the wounds in the early gestation fetus and in the oral mucosa of mammals heal without scar [6] the transition from scarless to scarred healing occurs in humans during late weeks of gestation [7]. A recent study showed that dermal fibroblasts with a scarring phenotype when transplanted into oral mucosa ended up generating more scar-like connective tissue compared with oral mucosal fibroblasts transplanted into the dermis [8]. The oral mucosal fibroblasts were shown to possess a higher baseline production capacity of several ECM-associated proteins than the skin fibroblasts, except type III collagen, which could be possibly attributed to a more favorable wound healing in oral mucosa [9]. Healthy endometrium heals scarlessly and is suggestive of regenerative healing and can be paralleled to fetal-like scarless healing responses that are also seen in the buccal mucosa of the oral cavity. Endometrial repair involves highly orchestrated cross-stalk in stromal, epithelial, vascular, and immune cells and presents a remarkable epitome of healing that involves over 400 cycles of resolution of inflammation, angiogenesis, tissue remodeling, and formation of new tissue without any residual scarring [10]. Recently, neutrophil gelatinase-associated lipocalin, follistatin like-1, chemokine ligand-20, and secretory leukocyte protease inhibitor were identified as important signatures in menstrual fluid that were proposed to facilitate scar-free repair [11]. Endometrial stromal cells were shown to exhibit distinct phenotypic and immunomodulatory profiles and displayed lack of HLA class II that was proposed to drive their physiological roles in tissue repair and immune tolerance during pregnancy [12].

Psoriasis represents a unique form of "scarless-like or hyper-regenerative" wound healing marked by nonscarring, inflammatory, and hyperproliferative tissue repair responses. An amazing aspect about the psoriatic lesions is the fact that with appropriate therapy the complex skin lesions can be reverted back to healthy appearing skin, with little if any evidence of altered changes in the epidermis and dermis. Psoriatic plaques are exciting conundrums as they do not go to fibrosis even amidst heralding auto-inflammatory loops [13]. An interesting common mediator oncofetal fibronectin extra domain B (Fn-EDB), has been reported to be prominent in psoriatic lesions and wound healing in fetal tissue [14]. Interestingly, psoriatic plaques despite being vulnerable to infections, do not tend to get infected because of the presence of massive antimicrobial peptides like LL37. Psoriasis pathogenesis involves strong polymorphonuclear neutrophil (PMN) infiltration and high levels of the PMN associated antimicrobial peptide, LL37. Psoriasis is marked by self-reactive inflammatory loops of innate immune responses, which trigger subsequent adaptive immune responses against auto-antigens like LL-37, ADAMTSL5, and HNRNPA1, with LL-37 and HNRNPA1 having RNA-binding properties. The phenomenon of self-RNA sensing by nucleic acid sensors [15–17] is central to autoinflammatory and autoimmune diseases like psoriasis, however, the role of RNA-binding proteins LL37 and HNRNPA1 (the proven autoantigens) in contributing to inflammatory loops remains largely unexplored. In a psoriatic mice model study excessive polyamine generation was shown to facilitate self-RNA sensing by immune cells [18] independent of RBPs, however, a recent study has implicated the role of neutrophil extracellular trap (NET)-associated RNA and LL37 (RBP) in self-amplifying inflammation in psoriasis [19]. Herster et al., highlight an unappreciated yet potential axis involving neutrophils, LL37 (RBP-like) and surprisingly, RNA that are abundant in psoriatic as opposed to healthy skin; suggesting a novel role of NET-derived RNA-RBP (LL37) complexes in self-propagating inflammatory loops. Host defence peptides or antimicrobial peptides like LL37 can have immunomodulatory protective [20] or pathological roles. The dual roles are proposed to be linked to post-translational modifications of peptides by citrullination or carbamylation that may depend on the disease context and result in altered ability of antimicrobial peptides to bind nucleic acids, thereby compromising their immunomodulatory potential (reviewed in [21]). Since RNA binding proteins (like LL37 and HNRNPA1) are emerging as potential molecules that can rewire inflammatory circuits depending on the pathological context, and several RBPs are also known to regulate developmental and fibrotic pathways by interacting with spliceosome machinery and acting as trans- regulators of RNA processing machinery [22], we would like to discuss their role in driving ECM remodeling in context of cardiac fibrosis with particular focus on RBM20, a cardiac-specific RBP that is emerging as a global regulator of cardiac development and disease.

2. Cardiac stress induced wound healing, repair and fibrosis: patch up, break up or a stressed extra-matrix affair

Plethora of extrinsic or intrinsic stressors (persistent hypertension, myocardial infarction, neurohormonal deregulation, hypoxia, ischemia-reperfusion, pressure over load, drug toxicity, mechanical stretch, radiation, etc.) can result in fibrosis and heart failure. The cardiac myocardium is a mosaic of diverse cell types with overrepresentation of cardiomyocytes and fibroblasts, and moderately populating endothelial cells, vessel smooth muscle cells, and immune cells like tissue resident macrophages. The absolute proportions of the cellular components are not determined; however, lineage tracing studies have identified the above inmates in the myocardium of healthy heart. Cardiac wound healing following a stress-induced injury (resulting in development of contraction bands, mitochondrial calcification and membrane disruption) involves a sequalae of pro-inflammatory, anti-inflammatory, and reparative events and majorly resulting in cardiomyocyte death and functional decline. Concomitantly, the neighboring fibroblasts in the myocardial niche act as central nodes to drive aberrant wound healing, ECM (extracellular matrix) remodeling and fibrosis. The continuum of tissue homeostasis, repair, and fibrosis is not appreciably understood, however, the inherent plasticity of fibroblasts and existence of pro- and anti- inflammatory, and pro-fibrogenic polarizing fibroblast phenotypes suggests a cardinal role of these cells in cardiac regeneration and repair. The most challenging aspect of cardiac remodeling is the recreation and restoration of qualitatively (structurally and functionally), and quantitatively compliant ECM, which is pro-regenerative or anti-fibrotic. It is known that extensive ECM deposition in the myocardial infarct can result in arrhythmias, however repopulating the native ECM following destruction is an absolute requirement for maintaining optimal tension in a highly contractile organ like heart. Fibrosis is projected to be the major player behind compromised myocardial compliance resulting in altered contraction-coupling events, reduced ventricular filling, decreased cardiac output and arrhythmias. End-stage heart failure and cardiac arrhythmia happen to associate with fibrosis, which is triggered as a compensatory response to counteract tissue damage, however, perpetuating cycles of stress and inflammation may result in decompensatory fibrosis and organ-failure during late-stage pathologies marked by recalcitrant accentuating cellular stress and catenating chronic inflammatory loops. Cardiac fibrosis is common in several cardiac diseases including atrial fibrillation, hypertrophic cardiomyopathy (HCM), dilated cardiomyopathy (DCM), and heart failure with preserved ejection fraction. Aberrant extracellular matrix remodeling can be present in myocardial ischemia and infarction as seen in ischemic (imbalanced oxygen supply and demand) heart diseases caused by atherosclerosis of the epicardial coronary arteries, and nonischemic heart diseases like aortic stenosis, diabetic cardiomyopathy, hypertensive heart disease, and hypertrophic cardiomyopathy, in which myocardial interstitial fibrosis results in adverse ventricular remodeling. Hypertrophying of ventricular cardiomyocytes is an initial adaptive response to compensate for cardiac over load and restore normalization of cardiac output by balancing ventricular wall stress, however, persistence of stress/stressors eventually culminates in cardiomyocyte death, fibroblast activation into myofibroblast, deposition of aberrant ECM, interstitial fibrosis, and adverse cardiac remodeling [23, 24].

Fibroblasts are key players in the secretion, deposition, organization and regulation of ECM turnover, however, their phenotypic heterogeneity, functional diversity, and attendant signalling pathways that modulate fibrotic over regenerative repair in cardiac diseases remain largely indecisive [25–37]. The sphingosine 1-phosphate (S1P) signaling pathway is a hot spot in research for fibrotic diseases involving lung, liver, and heart [29], and improved understanding of its cross-talk with fibroblasts is in progress as the attendant pathways are still controversial. Therefore, improved understanding of the molecular mechanisms and cross-talk underlying fibroblast activation and cardiomyocyte death are central for restoring cardiac homeostasis, designing novel regenerative approaches and developing anti-fibrotic therapies. Given the close association underlying tissue repair and fibrosis, it is intuitive that the balanced participation of common mediators can drive regeneration as opposed to skewed responses that may result in fibrosis. However, what are these checks and balances and how do they determine differential outcomes (homeostasis, repair or fibrosis) is the major challenge in the field of treating fibrosis or regenerating cardiac tissues *in vitro*. Delineating the subtle mechanisms and cellular, molecular, and extracellular matrix players implicated in regenerative and non-regenerative hearts can provide insights into the homeostasis-repair-fibrosis continuum, which remains the most vexing challenge in developing successful regenerative/anti-fibrotic approaches for fibrotic disorders. Further, the refractory response of cardiomyocytes to complete cell-cycle progression through mitosis limits their self-renewal, therefore, cardiomyogenic approaches to treat heart failure remain practically intractable [38–48].

3. Abortive cellular homeostasis rebalancing in cardiac fibrosis: a tragedy behind the broken heart

Cardiomyocytes being post-mitotic senesce with age and create cellular stress induced pool of biological waste over the course of ageing favoring late onset of fibrotic cardiomyopathies (acquired and inherited). Therefore, ageing and pro-fibrotic gene inheritance serve as additional "self-contained spontaneous stressors" that can impair the cellular homeostatic autophagic machinery, which is indispensable for cardiac tissue repair and proteostasis rebalancing [49–52]. Regulation of autophagy pathways are strongly implicated in liver, lung, heart, kidney and cystic fibrosis, which suggests that autophagy is a potential target for the treatment of chronic multiorgan fibrotic diseases involving aberrant extracellular remodeling [53]. The divergent homeostatic pathways converge to counter cellular stress or clear the stressor by intersecting and coordinating with precision, therefore, the intersection points, interacting partners and regulatory nodes are under extensive research. Molecular and cellular players driving homeostasis rebalancing in diverse diseases including fibrosis are continuing to emerge and recently reported for endoplasmic and mitochondrial stress pathway networks [54–56]. Mitochondria and endoplasmic reticulum (ER) in cardiomyocytes are overrepresented organelles to co-ordinate the increased metabolic demands and maintain active calcium stores for smooth flow. Endoplasmic reticulum and mitochondria quality control circuits are also integral to cardiac function, and deregulation of these pathways are strongly implicated in cardiac diseases including heart failure [57–63]. ECM remodeling is emerging to coincide with metabolic rewiring in cardiomyocytes and matrix-guided control of mitochondrial function in cardiomyocytes is seen as a potential therapeutic target in cardiac fibrosis, repair, regeneration and tissue engineering [64]. A recent trend in increasing ribosome profiling studies has stratified quality control checks that provide additional fidelity by clearance of defective messenger RNAs under ribosome associated quality control [65]. A genetic locus for cardiac hypertrophy that has been associated with alterations in stoichiometric translation rates of sarcomeric proteins has recently been defined [66]. Pro-fibrotic translationally regulated genes underlying cardiac fibrosis with proline (amino acid rich in collagen) codon usage promoted collagen synthesis emphasizing the importance of translational rates that may be heightened in failing hearts that select for codon biasing for profibrotic genes [67].

Importantly, the stress pathways gradually converge, overlap and cross-talk with RNA metabolic, and sterile inflammatory pathways through processing, secretion, and regulation of DAMPS (danger associated molecular patterns) generated in response to heralding stress and inflammation. DAMPs include diverse endogenous host-derived molecules (extracellular ATP, histones, HMGB1 chaperons, etc.), which can be sensed by innate immune receptors [17] owing to their cellular/extracellular mis-localization, stress induced modification or overexpression related conformational anomalies. DAMPs are primarily released by damaged and dying cells to facilitate sterile inflammation, which is important for tissue repair and regeneration, however, if left unchecked they can result in of numerous inflammatory diseases, metabolic disorders, neurodegenerative diseases, autoimmune diseases, cancer and fibrosis. Recently TGF-β has been proposed as an inducible DAMP that activates mechanotransducing pathways resulting in self-perpetuating loops leading to activation of myofibroblasts in diverse pathologies including cardiac fibrosis. Sarcomere integrity and rhythmic efficiency amidst high protein turnover, multiple protein-protein interactions, cyclic contractions and relaxations, and diverse "stress-stimuli" such as pressure overload, metabolic alterations, oxidative stress, hypoxia, ischemia-reperfusion, mechanical stress etc. is remarkable. However, it is also highly vulnerable to succumbing to these multiple stressors that can result in generation of DAMPs leading to inflammation, fibrosis and heart failure. The myocardium cell types express DAMP- sensing receptors and are proficient to respond immediately to stress and damage. Therefore, efficient quality control mechanism regulating cardiac homeostasis are indispensable to sarcomere maintenance and dynamic adaptation to stress [68–72].

4. Inflammatory networking in cardiac fibrosis: a heart on (in) flame

It is becoming increasingly recognized that the regenerative ability is not completely reliant on genetic makeup, environmental conditions or evolutionary hierarchies but the nature and extent of the immune responses to cardiac injury equally play important role in governing regenerative and non-regenerative modes of wound healing [73]. Importantly, cellular stress pathways cross- talk with inflammatory pathways that actively participate in restoring tissue homeostasis and overactivation of the inflammatory mediators can result in cardiomyocyte death and fibrosis. Macrophages are crucial for tissue homeostasis, following injury, circulating monocytes give rise to proinflammatory macrophages through activation mediated by DAMPs or cytokine secretion. Macrophages may contribute to cardiac fibrotic remodelling through secretion of TNF-α, IL-1β, IL-10, TGF-β and growth factors [74]. Studies have shown that Gata6 expressing macrophages can regulate cardiac fibrosis [75], CX3CR1$^+$ and CCR2$^+$ resident macrophages may positively or negatively regulate cardiac fibrosis, respectively following injury [76, 77]. Mast cells exist in low density in heart tissue, however, following an injury mast cells infiltrate heart tissue [78]. DAMPs may trigger degranulation of mast cells that leads to the release of inflammatory mediators including tryptase, chymase, TNF-α, and IL-1β [79]. Tryptase and chymase activate a potent fibrogenic mediator i.e. TGF-β (that promote myofibroblast differentiation and collagen production). Mast cells also produce PDGF-A and FGF2, which positively regulate fibrosis [80, 81]. However, studies have shown that mast cells can also produce IL-10 (anti-inflammatory agent) that is a negative regulator of fibrosis [82, 83]. Dendritic cells (DCs) play important role in initiating an adaptive immune response in post-injury cardiac remodelling. Studies have shown that DCs infiltrate cardiac tissue following an injury, specifically CD11+ DCs (bone marrow derived) are held crucial for cardiac homeostasis. Deficiency of CD11+ DCs following a cardiac injury may result in enhanced fibrosis [84]. Another study had shown that deletion of cardiac CD103+ DCs resulted in increased fibrosis [85]. Adaptive T and B lymphocytes are also central to cardiac inflammation as B and T cells infiltrate cardiac tissue following injury. There are different subsets of T cells including: CD4+, CD8+, CD73+ and Tregs. CD4+ cells have been reported to produce proinflammatory and fibrotic cytokines like IFN-α, following injury [86], while CD73 expressing T cells reduced fibrosis [87]. B cells secreted proinflammatory cytokines like IL-1β, IL-6 and TNF have been positively associated with fibrosis [88]. Neutrophils are known to regulate fibrosis in context dependent manner, however, neutrophil-derived extracellular traps (NETs) are becoming increasingly implicated in fibrotic pathologies including cardiac fibrosis. NETs have recently gained attention in chronic inflammatory, autoimmune and fibrotic settings including cystic fibrosis, interstitial lung disease, thromboinflammation, hypertrophic cardiomyopathy and liver fibrosis (reviewed in [89–95]). Notably, NETs have been reported to be associated with *disease-specific* bioactive proteins loaded onto them [96]. Intriguingly, emerging clinical and experimental studies indicate that neutrophils are able to release intrinsically and qualitatively different NETs decorated with *disease-specific* bioactive proteins dictated by diseased inflammatory environment containing tissue factor, IL-1β, IL-17, and LL37, suggesting systemic inflammation driven transcriptional-reprogramming in circulating neutrophils, which triggers *de novo* expression of disease-specific protein fingerprints that are extracellularly delivered through generation of NETs [97] and references therein, these exciting findings implicate NETs as potential anti-fibrotic targets. The non-immune cells of myocardial niche also participate in inflammatory responses, e.g. cardiomyocytes can generate pro-inflammatory mediators leading to profibrotic TGF-β and IGF-1 signalling [98, 99]. Endothelial cells can serve as both positive and negative regulator of fibrosis by generating profibrotic mediators like TGF-β, FGFs, or endothelin-1 [100] and undergoing endothelial to mesenchymal transition [101]. Endothelial cells express HIF-1 (hypoxia inducible factor) that can have anti-fibrotic effects [102] endothelial CXC chemokine Interferon-gamma-inducible protein (IP)-10/CXCL10, is also an anti-fibrotic molecule [103, 104].

The key observations that reflected elevated circulating proinflammatory cytokines in heart failure with reduced ejection fractions pumped the research into exploring role of immune system in heart failure pathogenesis. If inflammation is the cause or result of heart failure is still debatable, however, the developments in understanding the roles of innate and adaptive immune cells in heart failure are in active progress to identify heart failure patients who can have a cardio-inflammatory phenotype and can receive prospective anti-inflammatory and immunomodulatory regimens. The CANTOS trial with anti- IL-1β antibody canakinumab indicated decreased hospitalization rates in certain group of heart failure patients [105], these findings have renewed the interest in decoding cardio-inflammatory pathways for therapeutic targeting. Sensing of DAMPs can trigger non-cellular and cellular effectors in including IL-1, IL-6, IL-8, TNF, chemokines, complement system, inflammasome assembly, and activation of neutrophils, monocytes, macrophage innate immune cells that further engage the adaptive immune arm to trigger inflammatory loops [106]. Leukocyte dependent regulation of cardiac fibrosis is an ongoing area; however, it stays controversial and warrants further studies to exploit leukocyte plasticity and heterogeneity in cardiac fibrosis therapeutics [107]. Recent demonstration of engineered T cells or the CAR T-cell therapy directed against activated fibroblast specific antigen has sparked new hopes to existing limited clinical

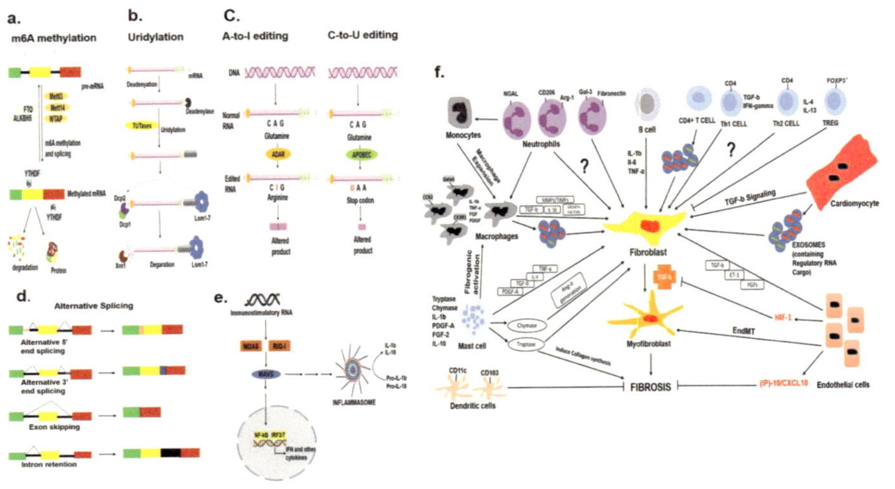

Figure 1.
RNA processing and modifications and their link to inflammation in cardiac fibrosis: (a) m6A Methylation, (b) Uridylation, (c) Editing, (d) Alternative Splicing, (e) Immunostimulatory RNA may result in Interferon and cytokines production. (f) Interaction between immune and non-immune cells during fibrotic remodeling. Diverse RNA processing pathways can result in generation of immunostimulatory RNAs that can trigger inflammatory cascade.

interventions and therapies in fibrotic heart failure [108, 109]. A recent study has demonstrated that macrophages expressing Mertk immune receptor in the heart supports cardiomyocyte health by phagocytosing exopher particles ejected from stressed cardiomyocytes harboring defective mitochondria. Mertk facilitated defective elimination of mitochondria from the myocardial tissue and prevented activation of the inflammasome, autophagy, metabolic stress, and ventricular dysfunction [110]. IL-11 signaling is also implicated in cardiac and cardiorenal fibrosis, however further studies will better indicate its precise role in driving pathogenesis [111]. A cross-talk between RNA processing pathways and immune- and non-immune cells through diverse mediators in context of cardiac fibrosis has been depicted in **Figure 1**.

5. Messenger RNA regulatory networks may modulate cellular stress and inflammation driving cardiac fibrosis: a message to the heart still in the outbox

The global co- and post- transcriptional mechanisms implicated in cardiac fibrosis are not well established, however, emerging studies are geared at extracting the subtle communications to identify intersection points between the cellular stress regulating pathways, regulatory non-coding RNAs, RNA metabolism intermediates/mediators and cardiac RNA binding proteins (RBFox, HuR, MBNL2, PUM2, QKI, CELF1, MBNL1, PTBP1) in context of cardiac diseases including fibrosis [66, 112–120]. A recent study implicated NUP155 subdomain hotspot with enriched allelic variants of the gene that suggests important role of RNA metabolism in cardiac disease and development [121].

Stem cell based regenerative approaches have found limited applicability in clinical translation to treat fibrosis. This ignited the research targeted at identifying cell- free secretory molecules that could not only have potential anti-fibrotic/regenerative potential to ameliorate fibrosis but characterization of these molecular players would also facilitate understanding of mechanisms underlying pathogenesis of fibrosis. In this context, bioactive vesicles/exosomes have been extensively investigated to explore the role of regulatory RNAs (generated through RNA processing pathways) in delivering pro- or anti- fibrotic outcomes, besides non-coding RNAs are also extensively studied in context of fibrosis independent of their exosomal/vesicular loading [120–141]. The cardiac inflammatory circuits also tend to converge at the regulatory networks controlled by RNA processing, metabolism and surveillance pathways. Particularly, the post- transcriptional regulation of cytokines to stabilize mRNA, determine the strength of proinflammatory pathways. Altered expression of AU rich (ARE) or GU rich (GRE) elements in cytokine and cytokine pathway intermediate transcripts impairs mRNA decay and can result in heightened immune responses as seen in diseased states [142]. Immunostimulatory RNAs (like double stranded RNAs) generated through deregulated RNA processing pathways along with RNA binding proteins of RNA helicase (RNA sensors) family are emerging as important components of immune response pathways during sterile inflammation that involves DAMP sensing [15, 16]. Mitochondrial quality control pathways intersecting with the endosomal compartments and lysosomes are recently reported to favor generation and release of mitochondrial-derived vesicles in former condition [143], further offering discernible biologically stable lipid vesicles that may help investigate how secreted cargos can impact tissue repair and homeostasis or trigger fibrosis, and can be extended to establishment of liquid biopsies for studying the progression of fibrotic diseases or alternatively these vesicles may serve as therapeutic tools like exosome-derived non coding RNAs [121, 125, 127, 128, 138, 139, 141, 144] circulating microRNAs (miRNAs) and tissue resident miRNAs play paradoxical role both as anti-fibrotic [145–148] and pro-fibrotic [149–151].

6. Harnessing RNA metabolic pathways in cardiac development and disease: getting to core of the heart

The cardiac output is tightly tuned to the functional outputs of the cardiac transcriptome or faithful expression of cardiac-specific genes. It is becoming evident that cellular processes (that are linked to generation of RNA variants) including alternative splicing, RNA editing, epitranscriptomic modifications like methylation, and alternative polyadenylation [152] have a major role in shaping the cardiac adaptive responses [119]. The advent of high throughput NGS sequencing has revealed striking diversity in RNA species/variants/isoforms that have revolutionized the field of RNA biology by informing on the codes of burgeoning RNA inventory, which has now been exploited in context of functional relevance of the neo-RNA entities in context of physiological and pathological outcomes. Expanding information and identification of genetic markers for heterogeneous complex diseases like heart failure has made it appreciably evident that cardiac development and differentiation cues are under tight regulation of splicing events [153], and mis-splicing of certain genes like titin (TTN) that is implicated in contractility and mechanosensation can result in adverse cardiac extracellular remodeling and fibrosis. Interestingly, single cell RNA sequencing of cochlear hair cells recently documented unappreciated complexity in splicing diversity and isoform abundance underlying biology of hearing and deafness, and reported sorcin (a key player in cardiac excitation-contraction) as a top hit in cochlear outer hair cells [154]. These exciting findings reflect potential shared mechanosensory targets that could result in co-manifestation of heterogenous genetic disorders like heart failure and hearing loss, which involve electrical conductance and contraction events (it remains to be explored if sorcin may have isoformic pattern of expression). It is already known that cardiac arrhythmias are a feature of Jervell and Lange-Nielsen syndrome (JLNS), an autosomal recessive disorder associated with congenital profound sensorineural hearing loss arising from homozygous or compound heterozygous mutations in either KCNQ1 or KCNE1 subunits of potassium ion channel conducting the slow component of the delayed rectifier current [155, 156].

Constitutive RNA splicing primarily involves the spliceosome machinery acting at the splice sites, however, alternative splicing may differ in its mechanism of action by engaging further cis- elements (regions within pre mRNA besides 5'/3' splice sites) like enhancers of exon/intron splicing [157]; the cis-elements are further acted upon by trans- acting modulators that constitute a family of RNA binding proteins (RBPs) with RNA binding motif (RBM), which further regulate negatively (repress) or positively (activate) the splice site selection. The activators include serine/arginine-domain–containing (SR) proteins and SR-related factors that dictate binding and assembly of the spliceosome complex and decide differential inclusion of exons in the mature transcript [158]. In contrast, the repressor trans-elements include the family of heterogenous nuclear ribonucleoproteins (hnRNPs) that tend to suppress splice site recognition [159, 160]. Therefore, alternative splicing events diversify the overall repository of functional and/or regulatory genes by inclusion/exclusion of exon, intron retention, alternative 5' or 3' splicing, and mutually exclusive exon utilization [119, 152], and at the same time splicing diversity may reflect pathological vulnerabilities dictated by the inherited variants that offer isoformic switching.

7. RBM20 interactome in cardiac development and disease: determining the soft- or hard- heartedness

Cardiac diseases including cardiomyopathy and arrhythmia are long known to be regulated by isoformic pattern of protein expression for genes including titin [161], CAMK2D, LDB3 and CACNA1C [162, 163]. Titin isoform-switching mechanisms at RNA (alternative splicing) and protein (post-translational modification) levels, which direct titin-based passive tension tuning remained largely elusive [164–169]. RNA binding protein RBM20, which is a splicing related factor was known to steer various aspects of cardiac function by regulating genes involved in biomechanics (TTN and TPM1), ion homeostasis and electrical activity (CAMK2D and CACNA1C) and signal transduction (CAMK2D and SPEN). Titin is the best exemplified target of RBM20 and TTN mutations are vastly implicated in cardiomyopathy [170, 171], and cardiovascular diseases [164, 172–174]. However, the mechanisms underlying alternative splicing of titin and role of thyroid hormone and insulin signaling in regulating it were nearly correlative [175–178] as the splice factors regulating alternative splicing in titin remained undetermined until early this decade. The role of post-transcriptional regulation in cardiac function and pathogenesis of human heart failure gained impetus following a pioneer study [179] on RNA binding motif (RBM20) protein, which has now emerged as a global regulator of cardiac alternative splicing isoformic switch in protein titin. RBM20 was found to be predominantly expressed in striated muscle, with maximum expression in the heart and its deficiency in rats was reported to resemble the pathophysiology of genetic dilated cardiomyopathy (fibrotic remodeling in heart). The RBM20-null rats exhibited increased subendocardial fibrosis with age and this effect was accompanied by electrical abnormalities and sudden death. The reduced activity of RBM20 (due to mutations/variants) resulted in altered isoform expression of genes central to biomechanics, electrophysiology and signal transduction culminating in cardiomyopathy, fibrosis, arrhythmia and sudden death [179]. Following these seminal findings that delineated the role of RBM20 driven splicing in titin isoforms, the ribonucleoprotein RBM20 has now paralled the role of titin in regulating structural and functional characteristics of cardiac development and disease. Cardiac-specific splicing events are attributed to RBM20, and recent studies with defective RBM20 variants have been shown to be associated with cardiac transcript variants resulting in cardiomyopathy including DCM (dilated cardiomyopathy) [180–190].

Structurally, RBM20 is a 1227 amino acid long protein with a leucine-rich N-terminal domain, zinc finger (ZnF) domain 1, RNA recognition motif (RRM) followed by mutational hotspot arginine-serine (RS) rich domain, glutamate (E) rich and ZnF2 regions located towards C-terminal. RRM and RS regions, and phosphorylation within RS region, are reported to be crucial for nuclear localization [179, 191, 192], RRM is important for binding to "UCUU" RNA sequence that dictates RBM20 binding to target genes. Several mutations are reported in RS region (predominantly the RSRSP stretch, amino acids 634-638) that likely disrupt its nuclear localization and hence splicing of target genes like titin, subsequently resulting in adverse cardiac modelling and familial dilated cardiomyopathy with associated fibrosis [148, 183, 193–203]. Compared to nuclear localization regions, the structural components contributing to splicing activity of RBM20 towards its targets are not fully explored, the near C-terminal E region has shown to have some contributions to splicing though [179, 204]. RBM20 is known to interact with spliceosome complex subunits U1 and U2 (small nuclear ribonucleic particles) and U2 related proteins like U2AF6 and U2AF35 [205]. The inventory of RBM20 regulated cardiac pre-mRNAs is dynamic and includes the following validated genes in human and rat- titin (TTN), calcium voltage gated channel subunit α1 C (CACNA1C), calcium/calmodulin dependent protein kinase II delta and gamma (CAMK2D and CAMK2G), formin homology 2 domain containing 3 (FHOD3), Lim domain binding 3 (LDB3), Lim domain only protein 7 (LMO7), muscular-enriched A type laminin-interacting protein (MLIP), PDZ and LIM domain 3 (PDLIM3), reticulon 4 (RTN4), ryanodine receptor 2 (RyR2), SH3 domain containing kinase binding protein 1 (SH3KBP1), sorbin and SH3 domain containing protein (SORBS1), and triadin (TRDN) [153, 179, 205–208].

Titin is an integral sarcomere protein responsible for maintaining passive elasticity in heart, structurally it is organized into modular structure with immunoglobulin-like (Ig), fibronectin type III (FnIII), proline (P), glutamate (E), valine (V), and lysine (K) containing highly elastic I-band region. The N-terminal domain of titin anchors it to the sarcomeric Z-band and C-terminal domain embeds it into M-band. The A-band maintains rigidity during contraction by binding to myosin. Titin's structural integrity is central to normal cardiac function, maintaining passive tension and driving length-dependent activation/Frank-Starling effect. Besides providing mechanical properties, titin stretching also participates in cellular signaling that facilitates cardiomyocyte growth and might be implicated in chronic myocardium remodeling, hypertrophy and fibrosis. Cardiomyopathy patients with mutations in titin gene further demonstrate the contribution of titin to systolic and diastolic heart failure. Systolic dysfunction underlies dilated cardiomyopathy (dilation of left ventricle) and hypertrophic cardiomyopathy (myocardial hypertrophy and ventricular thickening). Diastolic dysfunction is a hallmark of restrictive cardiomyopathy with preserved contractile force, however, abnormal relaxation during diastole results in decreased cardiac output due to inappropriate ventricle filling. Truncation mutations in titin gene cause dilated cardiomyopathy through diverse pathways that involve haploinsufficiency, activation of mTOR energy sensor pathways and increased metabolic stress (recently reviewed in [209]). Human induced pluripotent stem cell (hiPSCs) culture models aimed at generating cardiomyocytes from titin mutation carrying patients depict disorganized sarcomeric array, contraction disability and impaired force generation, however, similar extent of sarcomeric damage and myofibril contraction impairment has not been recapitulated in human studies or biopsied cardiomyocytes from titin variant patients. Therefore, it is alternatively proposed that titin variants may operate through creating a metabolic stress that could impair cardiac function independent of mutation sites by altering RNA metabolism pathways triggering non-sense mRNA decay (NMD) of abnormal titin variants and development of DCM phenotype. The cardiac metabolism could switch to branched chain amino acid pathway in place of fatty acid metabolism, deregulation of mTOR and autophagy pathways [210–218].

RBM20 cardiomyopathy has high penetrance and correlates with increased rates of heart failure, arrhythmias, and sudden cardiac death, new insights into RBM20 cardiomyopathy are extensively discussed recently [190]. Given the large size of titin (near 300kb) it is known to undergo extensive splicing events and yield several titin isoforms with cardiac N2B and N2BA to be the best characterized. N2B (shorter and stiff isoform) and N2BA (longer and pliant isoform) are adult cardiac isoforms of titin that regulate passive stiffness in the heart, and this is attributed to their structural dissimilarity in the highly elastic I-band region. Alternative splicing variants of titin during cardiac development keeps selecting for the shorter and stiffer isoform N2B in course of fetal to adult cardiac development, and physiologically N2B is overexpressed as compared to pliant N2BA isoform. However, aberrant expression patterns of titin isoforms resulting in altered ratios of N2B and N2BA are associated with cardiac diseases including cardiomyopathy with fibrosis and heart failure. RBM20 is shown to facilitate exon skipping events thereby selecting for shorter and stiffer forms of titin over development. In animal models, RBM20 homozygous mutations show increased ratio of N2BA/N2B (mirroring DCM phenotype), induced expression of RBM20 in RBM null mice decreases this ratio, however, intermediate effects (titin length, passive tension, sarcomere length) are seen in heterozygous mutations, indicating quantitative modulations of RBM20 as potential therapeutic approach to treat cardiac diseases [153, 179, 183, 194, 219]. RBM20 mutations in human patients result in severe inherited early onset DCM, manifesting even early on in younger patients with sudden death [190, 220]. Patient-specific stem cell based hiPSC culture models or CRISPR/CAS9 gene editing tools have been exploited to measure the effect of RBM20 point mutations in cardiomyocytes and alterations in sarcomere length, calcium handling, electrical coupling have been reported. The human iPSCs containing RBM20 mutations offer great tractable and tunable system to model cardiomyopathy *in vitro* and investigate potential signaling pathways contributing to the pro-fibrotic phenotypes [189, 221, 222]. The paradigm-shift in RNA metabolism associated interactome has begun to offer new therapeutic windows by unravelling the novel RNA binding proteins and splicing factors in context of cardiac development and fibrotic cardiomyopathies [119]. Biogenesis of regulatory non-coding RNAs i.e. microRNA, long noncoding RNA, and circular RNA, and their role in cardiac fibrosis, and RBM20 mediated alternative splicing of titin pre-mRNA is shown in **Figure 2**.

We briefly discuss the inflammatory networks in upper airway fibrotic diseases like laryngotracheal stenosis and subglottic stenosis that are relatively less explored in terms of mechanisms of fibrotic pathways, however, these pathologies need special attention as they might affect increasing number of patients given the current COVID-19 pandemic. Recent reports show that COVID-19 critically ill patients need mechanical ventilation, and many of these patients who need prolonged ventilation need surgical tracheostomy that is implicated in development of upper airway fibrosis.

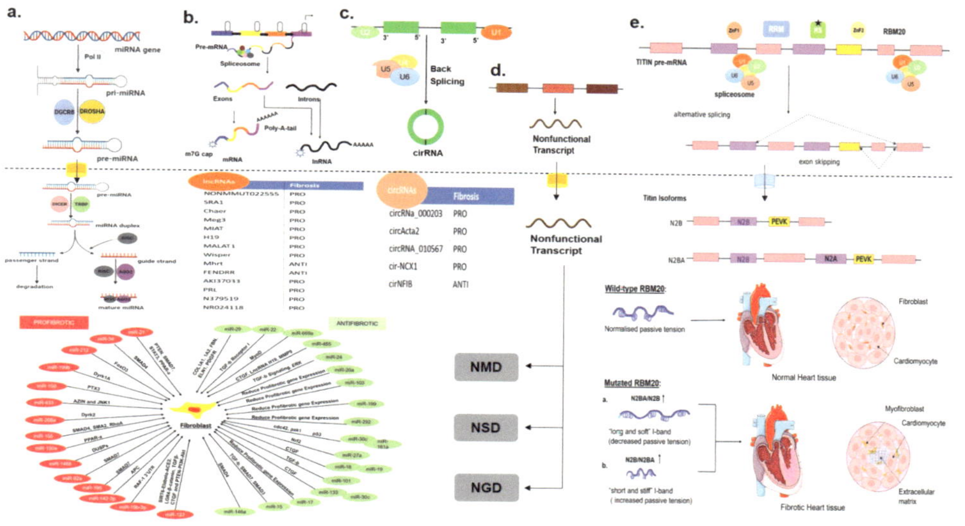

Figure 2.
Biogenesis of regulatory non-coding RNAs i.e. microRNA, long noncoding RNA and circular RNA and their role in cardiac fibrosis (a-c). Fate of aberrant RNA transcripts (d). RBM20 dependent splicing of Titin pre-mRNA resulting in formation of Titin isoforms which regulate cardiac development and fibrosis. Abbreviations: miRNA: microRNA; lnRNA: long non-coding RNA; cirRNA: circular RNA; NMD: nonsense-mediated decay; NSD: nonstop-mediated decay; NGD: no-go decay

8. Laryngotracheal stenosis: the pathogenesis and inflammatory pathways

Laryngotracheal stenosis (LTS) is an abnormal wound healing process of laryngotracheal mucosal inflammation, wound healing and scar formation. LTS is a fibrotic disease leading to pathologic narrowing of the larynx, subglottis, and trachea (the upper airway). There can be multiple etiologies to LTS, ranging from intubation injury (iatrogenic), radiation, autoimmune disease, to idiopathic [223]. The early stages of LTS are marked by dysphonia and communication difficulties that can develop into life-threatening progressive dyspnea leading to the airway compromise [224]. The most common form of LTS is the iatrogenic LTS (iLTS) caused by regional hypoxic and ischaemic pressure (stress) induced necrosis of the airway following prolonged intubation or tracheostomy [225]. The possible pathophysiology behind iatrogenic LTS is the surpassing of the pressure exerted by the cuff while prolonged intubations to that of the mucosal capillary perfusion pressure (approx. 35 mmHg), which results in ischemia, inflammation of the mucosa, and subsequent fibrotic strictures [226, 227]. The airway is primarily formed of 3 sets of cell types including epithelial cells, the fibroblasts, and the resident immune cells. The cross-talk of fibroblasts, immune cells and inflammatory cytokines participates in the development and propagation of LTS.

9. Inflammatory networks in the pathogenesis of LTS

TGFβ-SMAD2/3 cascade has been implicated in LTS and TGF-β antagonists have shown to attenuate fibrosis, however, TGF-β3 isoform has been reported to have antifibrotic response in LTS healing by significantly decreasing the inflammation and collagen deposition [228, 229], indicating opposing roles for isoforms of TGF-β. Hypoxia induced expression of IL-6 plays an important role in the pathogenesis of

LTS, the pressure exerted by the endotracheal tube cuff causes hypoxic and ischemic necrosis of the laryngotracheal mucosal tissue leading to inflammation and scarring marked by increased expression of IL-6, α-SMA & collagen, importantly IL-6 and myofibroblasts were also increased in an *ex-vivo* culture of healthy tracheal fibroblasts cultured under hypoxic conditions [230]. Possible role of B or T-cells in the formation of granulation tissue has also been been suggested [231]. Increased expression of profibrogenic Th2 cytokine IL-4 was seen in the brush biopsy samples of LTS scar and it stimulated fibroblast activation and excessive collagen formation in the LTS wound [225]. The expression of another Th2 cytokine IL-13 appeared to follow the same expression pattern of IL-4 and resulted in excessive fibrosis [232]. In contrast to Th2 cytokines, the Th1 cytokine IFN-γ inhibited fibrosis in LTS patients. Subsequent studies have shown the impact of IFN-γ on the LTS fibroblasts as significant decrease in levels of collagen and TGF-β expression was reported in the IFN-γ treated human LTS-derived fibroblasts compared to the untreated LTS-derived fibroblasts and normal laryngotracheal fibroblasts [226]. Dysregulated functioning of macrophages is related to fibroproliferative LTS [232], a prolonged cytokine signalling in the form of IL-4/IL-13 by Th2 cells can also contribute to impairment of macrophages by switching the non-fibroproliferative "classically activated" M1 macrophages into the fibroproliferative "alternatively activated" M2 macrophages [232]. Mice LTS model of chemical and mechanical injuries showed increased expression of M2 cell surface marker CD206 [224]. Inflammatory cytokine expression study in iLTS and autoimmune LTS patients demonstrated elevated levels of the macrophage growth factor granulocyte macrophage colony-stimulating factor (GM-CSF) and M2 cytokine IL-10 than that in controls [233]. Fibroblasts are the mesenchymal cells which are not terminally differentiated and rest in inactive state, under homeostasis. In their inactive but normal state they localise to the subepithelial layer of the airway tissue and provide for the biochemical and mechanical support to the tissue [234]. However, studies have reported increased ECM production and migration, and reduced contraction of iLTS fibroblasts [235], moreover, studies involving use of beta-aminopropionitrile (βAPN), an inhibitor of collagen cross-linking also demonstrated enhanced profibrotic features (overexpression of collagen I and II) in iLTS-derived fibroblasts [236], metabolically they exhibited enhanced glycolysis to oxidative phosphorylation ratio as see in proliferative cancer cells, justifying highly proliferative nature of iLTS-derived fibroblasts [237]. Emerging studies are reporting genetic link to LTS pathology, suggesting alternative treatment approaches to cure this fibrotic pathology. A functional single nucleotide polymorphism of TGF-β1 located in a negative regulatory element of its promoter was associated with the iatrogenic LTS. The data identified protective and susceptible genetic loci in patients undergoing endotracheal intubation. Another study, focused on 3 candidate genes encoding the innate immune receptor CD14, matrix metalloproteinase-1 (MMP-1), and the cytokine transforming growth factor β1 (TGF- β1). Reported association between MMP-1 and susceptibility to iLTS following intubation merits further investigation in a lager patient cohort [238].

10. Subglottic stenosis: a complex interplay between inflammation and fibrosis

Subglottic stenosis (SGS) is a relatively less explored fibrotic pathology in terms of mechanistic insights on inflammatory pathways. Researchers have gathered information to some extent by evaluating the changes that occur in the airway as a result of obliterative bronchiolitis [239], which shares the same fibrotic features of subglottic stenosis as studied in a murine model with emphasis on cytokines such as IL-1β, TGF-β and prostaglandin PGE2 [240]. SGS is accompanied by an acute and an exaggerated inflammatory response that triggers a shift in the cellular and molecular components in the healing wound in favor of more fibroblastic etiology [235]. SGS has numerous potential etiologies of which the most common cause in adults and children alike is prolonged endotracheal intubation. With developments in intensive care and associated intubations, it is natural that iatrogenic stenosis will become a major factor affecting post ICU quality of life. Animal and human studies have shown an upregulation of inflammatory markers in stenotic tissues. Patient factors like increased BMI, diabetes mellitus and chronic laryngopharyngeal reflux have also been implicated as causative factors for development of subglottic stenosis [241]. Cytokines like IL-1β, IL -10, TNF α, IFN γ and GM-CSF have shown significant increase in subglottic stenosis specimens [226]. Enhanced expression of profibrotic growth factors and cytokines like TGF-β, PDGF, IL-1, and Prostaglandin E2 was seen in patients of healing laryngeal lesions [242]. Expression of matrix metalloproteinases (MMPs), α-SMA, SMADs, IL-1 continued to rise 3 weeks beyond the initial insult [231]. Therefore, it appears that SGS undergoes an aberrant healing response as cytokines corresponding to various stages of wound healing process including inflammation (IL-1, TGFβ), proliferation (SMADs, TGFβ), and maturation (MMPs, αSMAs) are reported to present in pathogenesis of SGS. Idiopathic subglottic stenosis (iSGS) also has an inflammatory network evident from various studies suggesting the role of γδ T cells in IL-17A dependent tissue inflammation and airway remodeling in iSGS. Further studies delineating the role of RNA biology pathways might open up viable therapeutic options for this devastating pathology.

11. Future perspectives and summary

We are encouraged to chase the role of RBPs in homeostasis-tissue-repair-fibrosis continuum based on our recent preliminary findings, where we report for the first time an *in vitro* model that rigorously recapitulates proteolytic stress (as encountered in fibrotic pathologies) induced stress granule (SG) biomolecular condensate-like proteome signatures [243]. Dynamic phase separated membrane-less organelles including SGs are induced upon varied stress-stimuli (infectious or non-infectious) and are implicated in spatiotemporal control of various cellular functions including formation of signalling complexes, clustering of vesicles, sorting and trafficking of cargo [244]. Phase separated biomolecular condensates are becoming increasingly linked to developmental and pathological pathways [245–249]. We proposed proteases as novel stressors that can have diverse outcomes when present at varying concentrations (protease-antiprotease balance is crucial for driving tissue repair or fibrotic phenotypes). We observed heightened ribonucleoproteins (RNPs), spliceosome machinery, regulatory RNA generating proteins, and RNA binding proteins (RBPs) in our high-throughput proteomics data [243]. The formation of SGs like proteome was concomitant to translational halt in majority of proteins, sparing few essential cytoprotective proteins including exosome biogenesis and secretory pathway proteins that undergo synthesis despite stress environ. We hypothesize that the unique stress-associated proteins that represent "stress-essentialome" might get packaged and enriched into exosome vesicles in addition to the hitchhiking of SG regulatory RNAs, RBPs, RNPs and cytoprotective proteins onto the exosomal carriers leading to conglomeration of unique disease/stressor-specific cargos in exosome silos. Translational halt would result in polysome run off and dissociation of ribosome and translating mRNAs that would partition into the stress granules, therefore, RNA and polysome profiling of stressed cells in addition to exosome cargo profiling might offer valuable information on

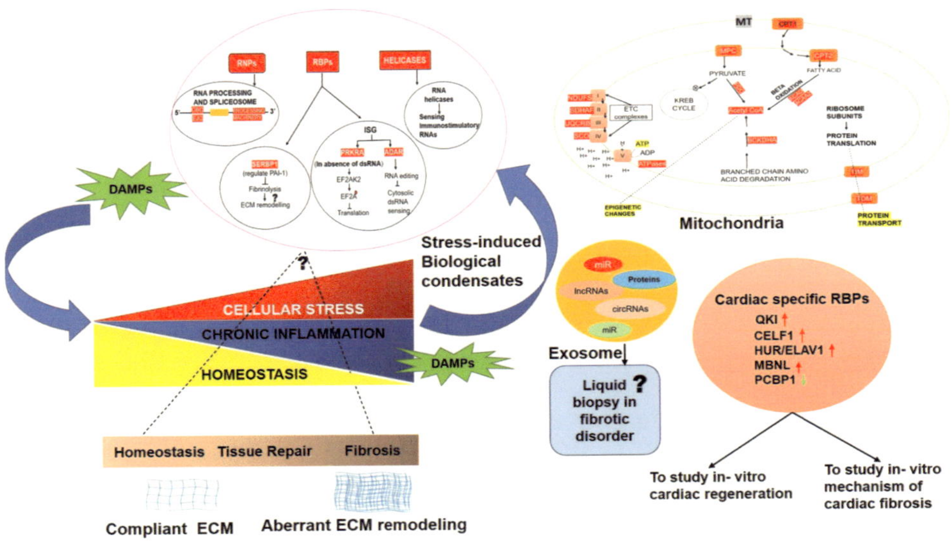

Figure 3.
Proposed model for cellular stress induced biological condensates that can regulate homeostasis, tissue repair and fibrosis. A cellular-stress induced reshaping of RNA processing machinery by generation of biomolecular condensates, and their coupling to mitochondrial and exosomal pathways. The stress-coupled exosomes are proposed to carry pathologic cargo which can be exploited to develop liquid biopsies in the context of fibrotic disorders. The expression of cardiac-specific RBPs in our model can be utilized to develop cardiac regenerative approaches in vitro or to study the role of RBPs in cardiac fibrosis. RNPs: ribonucleoproteins: RBPs: RNA binding proteins; CBC: cap binding complex; EJC: exon junction complex; SERBP1: SERPINE1 mRNA-binding protein 1; PAI-1: Plasminogen activator inhibitor-1; ISG: Interferon stimulated gene; PRKRA: Protein kinase, interferon-inducible double stranded RNA dependent activator; ADAR: Adenosine deaminases acting on RNA; EF2AK2: Eukaryotic Translation Initiation Factor 2-alpha Kinase 2; RED color: upregulated proteins; MT: mitochondrion; CPT: Carnitine palmitoyl-transferase; MPC: mitochondrial pyruvate carrier; PDC: Pyruvate dehydrogenase complex; ACAD: Acyl-CoA dehydrogenase; ACAA2: acetyl-Coenzyme A acyltransferase 2; BCKDHA: branched-chain alpha-keto acid dehydrogenase; NDUFS: NADH-ubiquinone oxidoreductase subunit; SDHAF: Succinate dehydrogenase complex assembly factor 1; UQCRB: Ubiquinol-cytochrome c reductase binding protein; SCO: synthesis of cytochrome c oxidase; TOM: translocase of the outer membrane; TIM: translocase of the inner membrane; ETC: electron transport chain. DAMPs- danger associated molecular patterns.

pathologic transcripts, associated regulatory RNAs (miRNAs, lncRNA) and translating ribosome composition besides the proteome signatures, thereby offering new dimensions to investigate stress-induced regenerative or fibrotic responses and the role of RNA processing machinery and RBPs in driving these triggers. It might offer identification of cell or tissue specific splicing factors and an opportunity to attempt rescuing splicing related alterations inherited in genome (pathogenic variants), to switch for native isoforms. In addition, stress-induced secretion of extracellular vesicles/exosomes could offer novel therapeutic opportunities by developing liquid biopsies in fibrotic diseases (a less explored area), which may yield meaningful information and help predict progression of the disease in suitable disease-specific *in vitro* models. Exosomes are relatively stable, can avoid background noise, can be easily detected from blood and their molecular analysis may decipher pathobiology of disease, fibrotic liver derived exosomes (in mice) showed increased CCN2; decreased Twist1, miR-214 than control mice [250]. Circulating exosomes from mice with alcohol related liver disease when transmitted to normal mice resulted in pro-inflammatory/fibrogenic liver phenotype [251]. MiR-125 has also been found to be upregulated in serum from patients with cirrhosis than controls [252]. Multiple studies *in-vitro* and in mice have revealed pathologic micro-RNAs associated with liver injury, liver fibrosis and liver malignancy [250]. However, major challenge in translating bench to bedside knowledge include reliable standardization and characterization protocols, and validation in a well-characterized patient population. Therefore, we are interested in deciphering the molecular information stored in exosomes of patients with liver cirrhosis using standardized protocol and their correlation with key events like death, sepsis, organ failures. Our proposed model of stress- induced reshaping of RNA metabolic pathways that are coupled to mitochondrial alterations and exosome biosynthesis is shown in **Figure 3**.

Acknowledgements

MS thanks financial support from PGIMER grant: No.71/2-Edu-16/1098, and SERB funded grants-SERB/CRG/2019/006745, and SPR/2019/001447. KK thanks SERB/CRG/2019/006745 for providing Junior Research Fellowship.

Author details

Maryada Sharma[1*], Kavita Kaushal[1], Sanjay Singh Rawat[1], Manjul Muraleedharan[1], Seema Chhabra[2], Nipun Verma[3], Anupam Mittal[4], Ajay Bahl[5], Madhu Khullar[6], Anurag Ramavat[1] and Naresh K. Panda[1]

1 Department of Otolaryngology and Head and Neck Surgery, Postgraduate Institute of Medical Education and Research, Chandigarh, India

2 Department of Immunopathology, Postgraduate Institute of Medical Education and Research, Chandigarh, India

3 Department of Hepatology, Postgraduate Institute of Medical Education and Research, Chandigarh, India

4 Department of Translational and Regenerative Medicine, Postgraduate Institute of Medical Education and Research, Chandigarh, India

5 Department of Cardiology, Postgraduate Institute of Medical Education and Research, Chandigarh, India

6 Department of Experimental Medicine and Biotechnology, Postgraduate Institute of Medical Education and Research, Chandigarh, India

*Address all correspondence to: maryada24@yahoo.com; sharma.maryada@pgimer.edu.in

References

[1] M. Sharma, A Tiwari, S. Sharma, P. Bhoria, V. Gupta, A. Gupta, and M. Luthra-Guptasarma (2013) Fibrotic remodeling of the extracellular matrix through a novel (engineered, dual-function) antibody reactive to a cryptic epitope on the N-terminal 30 kDa fragment of fibronectin. PLOS one 8 (7): e69343

[2] M. Sharma, A Tiwari, S. Sharma, R. Bansal, V. Gupta, A. Gupta, and M. Luthra-Guptasarma (2014) Pathologic vitreous causes cell line-derived (but not donor-derived) retinal pigment epithelial cells to display PVR-like features in culture. Clinical and Experimental Ophthalmology; doi: 10.1111/ceo.12307

[3] A. Tiwari, R. Kumar, J. Ram, M. Sharma* and M. Luthra-Guptasarma* (2016) Control of fibrotic changes through the synergistic effects of anti-fibronectin antibody and an RGDS-tagged form of the same antibody. *Scientific Reports 6:30872*

[4] Brij Bhushan Mehta, Anil Tiwari, Saniya Sharma, Ashu Shukla, *M. Sharma*, Rakesh K Vasishta, Ramesh Sen, Aman Sharma, Manni Luthra-Guptasarma (2018). Amelioration of collagen antibody induced arthritis in mice by an antibody directed against the fibronectin type III repeats of tenascin-C. *International Immunopharmacology* 58, 15-23

[5] Santra M, Sharma M, Katoch D, Jain S, Saikia UN, Dogra MR, Luthra-Guptasarma M. Induction of posterior vitreous detachment (PVD) by non-enzymatic reagents targeting vitreous collagen liquefaction as well as vitreoretinal adhesion. Sci Rep. 2020 Jul 16;10(1):12083. doi: 10.1038/s41598-020-69093-w.

[6] Marshall CD, Hu MS, Leavitt T, Barnes LA, Lorenz HP, Longaker MT. Cutaneous Scarring: Basic Science, Current Treatments, and Future Directions. *Adv Wound Care (New Rochelle)*. 2018;7(2):29-45. doi: 10.1089/wound.2016.0696kin.

[7] Karppinen SM, Heljasvaara R, Gullberg D, Tasanen K, Pihlajaniemi T. Toward understanding scarless skin wound healing and pathological scarring. *F1000Res*. 2019;8:F1000 Faculty Rev-787. Published 2019 Jun 5. doi:10.12688/f1000research.18293.1

[8] Nikoloudaki G, Creber K, Hamilton DW. Wound healing and fibrosis: a contrasting role for periostin in skin and the oral mucosa. Am J Physiol Cell Physiol. 2020 Jun 1;318(6):C1065-C1077. doi: 10.1152/ajpcell.00035.2020. Epub 2020 Apr 8. PMID: 32267719; PMCID: PMC7311745.

[9] Hara-Saito, Y., Kato, H., Saito, N. *et al.* Distinct differences in hypoxic responses between human oral mucosa and skin fibroblasts in a 3D collagen matrix. *In Vitro Cell Dev.Biol.-Animal* 2020; **56**, 452-479 (2020). https://doi.org/10.1007/s11626-020-00458-1

[10] Critchley HO, Maybin JA, Armstrong GM, Williams AR. Physiology of the Endometrium and Regulation of Menstruation. Physiological reviews. 2020 Jul 1;100(3):1149-79.

[11] Evans J, Infusini G, McGovern J, Cuttle L, Webb A, Nebl T, Milla L, Kimble R, Kempf M, Andrews CJ, Leavesley D. Menstrual fluid factors facilitate tissue repair: identification and functional action in endometrial and skin repair. The FASEB Journal. 2019 Jan;33(1):584-605.

[12] Queckboerner S, Von Grothusen C, Boggavarapu NR, Davies LC, Gemzell-Danielsson K. Stromal Heterogeneity in the Proliferative Endometrial

Functionalis-A single-cell approach. BioRxiv. 2020 Jan 1.

[13] Nickoloff BJ, Bonish BK, Marble DJ, Schriedel KA, DiPietro LA, Gordon KB, Lingen MW. Lessons learned from psoriatic plaques concerning mechanisms of tissue repair, remodeling, and inflammation. J Investig Dermatol Symp Proc. 2006 Sep;11(1):16-29. doi: 10.1038/sj.jidsymp.5650010. PMID: 17069007.

[14] Georgescu SR, Tampa M, Caruntu C, et al. Advances in Understanding the Immunological Pathways in Psoriasis. Int J Mol Sci.

[15] Liu G, Gack MU. Distinct and Orchestrated Functions of RNA Sensors in Innate Immunity. Immunity. 2020 Jul 14;53(1):26-42.

[16] McWhirter SM, Jefferies CA. Nucleic Acid Sensors as Therapeutic Targets for Human Disease. Immunity. 2020 Jul 14;53(1):78-97.

[17] Gong T, Liu L, Jiang W, Zhou R. DAMP-sensing receptors in sterile inflammation and inflammatory diseases. Nature Reviews Immunology. 2019 Sep 26:1-8.

[18] Lou F, Sun Y, Xu Z, Niu L, Wang Z, Deng S, Liu Z, Zhou H, Bai J, Yin Q, Cai X. Excessive Polyamine Generation in Keratinocytes Promotes Self-RNA Sensing by Dendritic Cells in Psoriasis. Immunity. 2020 Jul 14;53(1):204-16.

[19] Herster F, Bittner Z, Archer NK, Dickhöfer S, Eisel D, Eigenbrod T, Knorpp T, Schneiderhan-Marra N, Löffler MW, Kalbacher H, Vierbuchen T. Neutrophil extracellular trap-associated RNA and LL37 enable self-amplifying inflammation in psoriasis. Nature communications. 2020 Jan 8;11(1):1-3

[20] Alford MA, Baquir B, Santana FL, Haney EF, Hancock RE. Cathelicidin Host Defense Peptides and Inflammatory Signaling: Striking a Balance. Frontiers in Microbiology. 2020 Aug 27;11:1902.

[21] Liang W, Diana J. The Dual Role of Antimicrobial Peptides in Autoimmunity. Frontiers in Immunology. 2020 Sep 2;11:2077.

[22] Gao C, Wang Y. mRNA Metabolism in Cardiac Development and Disease: Life After Transcription. Physiological Reviews. 2020 Apr 1;100(2):673-94.

[23] Frangogiannis NG. Cardiac fibrosis: cell biological mechanisms, molecular pathways and therapeutic opportunities. Molecular aspects of medicine. 2019 Feb 1;65:70-99

[24] Frangogiannis NG, Kovacic JC. Extracellular Matrix in Ischemic Heart Disease, Part 4/4: JACC Focus Seminar. Journal of the American College of Cardiology. 2020 May 5;75(17):2219-35.

[25] Daseke II MJ, Tenkorang MA, Chalise U, Konfrst SR, Lindsey ML. Cardiac Fibroblast Activation during Myocardial Infarction Wound Healing: Fibroblast polarization after MI. Matrix Biology. 2020 May 21.

[26] Delaunay M, Osman H, Kaiser S, Diviani D. The Role of Cyclic AMP Signaling in Cardiac Fibrosis. Cells. 2020 Jan;9(1):69

[27] González A, López B, Ravassa S, San José G, Díez J. Reprint of "The complex dynamics of myocardial interstitial fibrosis in heart failure. Focus on collagen cross-linking". Biochimica et Biophysica Acta (BBA)-Molecular Cell Research. 2020 Mar 1;1867(3):118521.

[28] Kingma JG. Myocardial Infarction: Perspectives on Cardiac Regeneration and Cardiac Remote Conditioning Interventions to Limit Cellular Injury. World Journal of Cardiovascular Diseases. 2020 Apr 22;10(04):188.

[29] Wang E, He X, Zeng M. The role of S1P and the related signaling pathway in the development of tissue fibrosis. Frontiers in Pharmacology. 2019 Jan 8;9:1504.

[30] Ceccato TL, Starbuck RB, Hall JK, Walker CJ, Brown TE, Killgore JP, Anseth KS, Leinwand LA. Defining the Cardiac Fibroblast Secretome in a Fibrotic Microenvironment. Journal of the American Heart Association. 2020 Sep 12:e017025.

[31] Humeres C, Frangogiannis NG. Fibroblasts in the infarcted, remodeling, and failing heart. JACC: Basic to Translational Science. 2019 Jun 1;4(3):449-67.

[32] Nwabuo CC, Vasan RS. Pathophysiology of hypertensive heart disease: beyond left ventricular hypertrophy. Current Hypertension Reports. 2020 Feb 1;22(2):11.

[33] Wingard MC, Frasier CR, Singh M, Singh K. Heart failure and diabetes: role of ATM. Current Opinion in Pharmacology. 2020 Oct 1;54:27-35.

[34] Angelini A, Trial J, Ortiz-Urbina J, Cieslik KA. Mechanosensing dysregulation in the fibroblast: a hallmark of the aging heart. Ageing Research Reviews. 2020 Aug 23:101150.

[35] DeLeon-Pennell KY, Barker TH, Lindsey ML. Fibroblasts: The arbiters of extracellular matrix remodeling. Matrix Biology. 2020 Jun 3.

[36] Cojan-Minzat BO, Zlibut A, Agoston-Coldea L. Non-ischemic dilated cardiomyopathy and cardiac fibrosis. Heart Failure Reviews. 2020 Mar 13:1-21

[37] Czubryt MP. Cardiac Fibroblast to Myofibroblast Phenotype Conversion—An Unexploited Therapeutic Target. Journal of cardiovascular development and disease. 2019 Sep;6(3):28.

[38] Frangogiannis NG. Transforming growth factor–β in tissue fibrosis. Journal of Experimental Medicine. 2020 Mar 2;217(3).

[39] Hanna A, Frangogiannis NG. The role of the TGF-beta superfamily in myocardial infarction. Frontiers in cardiovascular medicine. 2019;6:140.

[40] Hortells L, Johansen AK, Yutzey KE. Cardiac Fibroblasts and the Extracellular Matrix in Regenerative and Nonregenerative Hearts. Journal of cardiovascular development and disease. 2019 Sep;6(3):29.

[41] Balbi C, Costa A, Barile L, Bollini S. Message in a bottle: upgrading cardiac repair into rejuvenation. Cells. 2020 Mar;9(3):724.

[42] Broughton KM, Sussman MA. Adult cardiomyocyte cell cycle detour: off-ramp to quiescent destinations. Trends in Endocrinology & Metabolism. 2019 Aug 1;30(8):557-67.

[43] Pagano F, Picchio V, Chimenti I, Sordano A, De Falco E, Peruzzi M, Miraldi F, Cavarretta E, Zoccai GB, Sciarretta S, Frati G. On the Road to Regeneration:"Tools" and "Routes" Towards Efficient Cardiac Cell Therapy for Ischemic Cardiomyopathy. Current cardiology reports. 2019 Nov 1;21(11):133.

[44] Dronkers E, Wauters MM, Goumans MJ, Smits AM. Epicardial TGFβ and BMP Signaling in Cardiac Regeneration: What Lesson Can We Learn from the Developing Heart?. Biomolecules. 2020 Mar;10(3):404.

[45] Marchianò S, Bertero A, Murry CE. Learn from your elders: developmental biology lessons to guide maturation of stem cell-derived cardiomyocytes.

Pediatric Cardiology. 2019 Oct 1;40(7):1367-87

[46] Leitolis A, Robert AW, Pereira IT, Correa A, Stimamiglio MA. Cardiomyogenesis modeling using pluripotent stem cells: the role of microenvironmental signaling. Frontiers in cell and developmental biology. 2019;7:164.

[47] Guo Y, Pu WT. Cardiomyocyte maturation: new phase in development. Circulation Research. 2020 Apr 10;126(8):1086-106.

[48] Israeli Y, Gabalski M, Ball K, Wasserman A, Zou J, Ni G, Zhou C, Aguirre A. Generation of Heart Organoids Modeling Early Human Cardiac Development Under Defined Conditions. Available at SSRN 3654622. 2020 Jan 1.

[49] Abdellatif M, Sedej S, Carmona-Gutierrez D, Madeo F, Kroemer G. Autophagy in cardiovascular aging. Circulation research. 2018 Sep 14;123(7):803-24.

[50] Abdellatif M, Ljubojevic-Holzer S, Madeo F, Sedej S. Autophagy in cardiovascular health and disease. Prog Mol Biol Transl Sci. 2020 May 12;172:87-106.

[51] Islas-Carbajal MC, Rincón-Sánchez AR, Nava-Valdivia CA, Charles-Niño CL. The Importance of Autophagy and Proteostasis in Metabolic Cardiomyopathy. InCardiovascular Risk Factors in Pathology 2020 Jun 18. IntechOpen.

[52] Packer M. Autophagy stimulation and intracellular sodium reduction as mediators of the cardioprotective effect of sodium–glucose cotransporter 2 inhibitors. European Journal of Heart Failure. 2020 Apr;22(4):618-28

[53] Li Y, Liu R, Wu J, Li X. Self-eating: friend or foe? The emerging role of autophagy in fibrotic diseases. Theranostics. 2020;10(18):7993.

[54] Zhao J, Qi YF, Yu YR. STAT3, a key regulator in liver fibrosis. Annals of Hepatology. 2020 Jul 21.

[55] Grandjean JM, Madhavan A, Cech L, Seguinot BO, Paxman RJ, Smith E, Scampavia L, Powers ET, Cooley CB, Plate L, Spicer TP. Pharmacologic IRE1/XBP1s activation confers targeted ER proteostasis reprogramming. Nature Chemical Biology. 2020 Jul 20:1-0.

[56] Chareyron I, Wall C, Thevenet J, Santo-Domingo J, Wiederkehr A. Cellular stress is a prerequisite for glucose-induced mitochondrial matrix alkalinization in pancreatic β-cells. Molecular and cellular endocrinology. 2019 Feb 5;481:71-83.

[57] Belmadani S, Matrougui K. Broken heart: A matter of the endoplasmic reticulum stress bad management?. World journal of cardiology. 2019 Jun 26;11(6):159.

[58] Quiles JM, Gustafsson ÅB. Mitochondrial Quality Control and Cellular Proteostasis: Two Sides of the Same Coin. Frontiers in Physiology. 2020;11.

[59] Ajoolabady A, Aslkhodapasandhokmabad H, Aghanejad A, Zhang Y, Ren J. Mitophagy Receptors and Mediators: Therapeutic Targets in the Management of Cardiovascular Ageing. Ageing Research Reviews. 2020 Jul 22:101129.

[60] Rogers RG, Ciullo A, Marbán E, Ibrahim AG. Extracellular Vesicles as Therapeutic Agents for Cardiac Fibrosis. Frontiers in Physiology. 2020;11:479.

[61] Zhu B, Zhang L, Liang C, Liu B, Pan X, Wang Y, Zhang Y, Zhang Y, Xie W, Yan B, Liu F. Stem cell-derived exosomes prevent aging-induced

cardiac dysfunction through a novel exosome/lncRNA MALAT1/NF-κB/TNF-α signaling pathway. Oxidative Medicine and Cellular Longevity. 2019 Apr 8;2019.

[62] Svaguša T, Martinić M, Martinić M, Kovačević L, Šepac A, Miličić D, Bulum J, Starčević B, Sirotković-Skerlev M, Seiwerth F, Kulić A. Mitochondrial unfolded protein response, mitophagy and other mitochondrial quality control mechanisms in heart disease and aged heart. Croatian Medical Journal. 2020 Apr;61(2):126.

[63] Lin R, Kerkelä R. Regulatory Mechanisms of Mitochondrial Function and Cardiac Aging. International Journal of Molecular Sciences. 2020 Jan;21(4):1359.

[64] Lyra-Leite DM, Andres AM, Cho N, Petersen AP, Ariyasinghe NR, Kim SS, Gottlieb RA, McCain ML. Matrix-guided control of mitochondrial function in cardiac myocytes. Acta biomaterialia. 2019 Oct 1;97:281-95.

[65] Hickey KL, Dickson K, Cogan JZ, Replogle JM, Schoof M, D'Orazio KN, Sinha NK, Hussmann JA, Jost M, Frost A, Green R. GIGYF2 and 4ehp inhibit translation initiation of defective messenger RNAs to assist Ribosome-Associated quality control. Molecular cell. 2020 Jul 28.

[66] Witte F, Ruiz-Orera J, Mattioli CC, Blachut S, Adami E, Schulz JF, Schneider-Lunitz V, Hummel O, Patone G, Mücke MB, Šilhavý J. Trans control of cardiac mRNA translation in a protein length-dependent fashion. bioRxiv. 2020 Jan 1.

[67] Wu J, Subbaiah KC, Xie LH, Jiang F, Khor ES, Mickelsen D, Myers JR, Tang WH, Yao P. Glutamyl-Prolyl-tRNA Synthetase Regulates Proline-Rich Pro-Fibrotic Protein Synthesis During Cardiac Fibrosis. Circulation Research. 2020 Aug 28;127(6):827-46.

[68] Zheng M, Jacob J, Hung SH, Wang J. The Hippo Pathway in Cardiac Regeneration and Homeostasis: New Perspectives for Cell-Free Therapy in the Injured Heart. Biomolecules. 2020 Jul;10(7):1024.

[69] Land WG. Use of DAMPs and SAMPs as Therapeutic Targets or Therapeutics: A Note of Caution. Molecular Diagnosis & Therapy. 2020 Apr 4:1.

[70] Seclì L, Sorge M, Morotti A, Brancaccio M. Blocking Extracellular Chaperones to Improve Cardiac Regeneration. Frontiers in Bioengineering and Biotechnology. 2020 May 26;8:411.

[71] Takahashi T, Shishido T, Kinoshita D, Watanabe K, Toshima T, Sugai T, Narumi T, Otaki Y, Tamura H, Nishiyama S, Arimoto T. Cardiac nuclear high-mobility group box 1 ameliorates pathological cardiac hypertrophy by inhibiting DNA damage response. JACC: Basic to Translational Science. 2019 Apr 29;4(2):234-47.

[72] Liu FY, Fan D, Yang Z, Tang N, Guo Z, Ma SQ, Ma ZG, Wu HM, Deng W, Tang QZ. TLR9 is essential for HMGB1-mediated post-myocardial infarction tissue repair through affecting apoptosis, cardiac healing, and angiogenesis. Cell death & disease. 2019 Jun 17;10(7):1-6.

[73] Lai SL, Marín-Juez R, Stainier DY. Immune responses in cardiac repair and regeneration: a comparative point of view. Cellular and Molecular Life Sciences. 2019 Apr 15;76(7):1365-80.

[74] Frangogiannis NG. Cardiac fibrosis: cell biological mechanisms, molecular pathways and therapeutic opportunities. Molecular aspects of medicine. 2019 Feb 1;65:70-99.

[75] Deniset JF, Belke D, Lee WY, Jorch SK, Deppermann C,

Hassanabad AF, Turnbull JD, Teng G, Rozich I, Hudspeth K, Kanno Y. Gata6+ pericardial cavity macrophages relocate to the injured heart and prevent cardiac fibrosis. Immunity. 2019 Jul 16;51(1):131-40.

[76] Liao X, Shen Y, Zhang R, Sugi K, Vasudevan NT, Alaiti MA, Sweet DR, Zhou L, Qing Y, Gerson SL, Fu C. Distinct roles of resident and nonresident macrophages in nonischemic cardiomyopathy. Proceedings of the National Academy of Sciences. 2018 May 15;115(20):E4661-9.

[77] Ma Y, Mouton AJ, Lindsey ML. Cardiac macrophage biology in the steady-state heart, the aging heart, and following myocardial infarction. Translational Research. 2018 Jan 1;191:15-28.

[78] Ngkelo A, Richart A, Kirk JA, Bonnin P, Vilar J, Lemitre M, Marck P, Branchereau M, Le Gall S, Renault N, Guerin C. Mast cells regulate myofilament calcium sensitization and heart function after myocardial infarction. Journal of Experimental Medicine. 2016 Jun 27;213(7):1353-74.

[79] Legere SA, Haidl ID, Légaré JF, Marshall JS. Mast cells in cardiac fibrosis: new insights suggest opportunities for intervention. Frontiers in immunology. 2019 Mar 28;10:580.

[80] Liao CH, Akazawa H, Tamagawa M, Ito K, Yasuda N, Kudo Y, Yamamoto R, Ozasa Y, Fujimoto M, Wang P, Nakauchi H. Cardiac mast cells cause atrial fibrillation through PDGF-A–mediated fibrosis in pressure-overloaded mouse hearts. The Journal of clinical investigation. 2010 Jan 4;120(1):242-53.

[81] Wernersson S, Pejler G. Mast cell secretory granules: armed for battle. Nature Reviews Immunology. 2014 Jul;14(7):478-94.

[82] Krishnamurthy P, Rajasingh J, Lambers E, Qin G, Losordo DW, Kishore R. IL-10 inhibits inflammation and attenuates left ventricular remodeling after myocardial infarction via activation of STAT3 and suppression of HuR. Circulation research. 2009 Jan 30;104(2):e9-18.

[83] Verma SK, Garikipati VN, Krishnamurthy P, Schumacher SM, Grisanti LA, Cimini M, Cheng Z, Khan M, Yue Y, Benedict C, Truongcao MM. Interleukin-10 inhibits bone marrow fibroblast progenitor cell–mediated cardiac fibrosis in pressure-overloaded myocardium. Circulation. 2017 Sep 5;136(10):940-53.

[84] Anzai A, Anzai T, Nagai S, Maekawa Y, Naito K, Kaneko H, Sugano Y, Takahashi T, Abe H, Mochizuki S, Sano M. Regulatory role of dendritic cells in postinfarction healing and left ventricular remodeling. Circulation. 2012 Mar 13;125(10):1234-45.

[85] Clemente-Casares X, Hosseinzadeh S, Barbu I, Dick SA, Macklin JA, Wang Y, Momen A, Kantores C, Aronoff L, Farno M, Lucas TM. A CD103+ conventional dendritic cell surveillance system prevents development of overt heart failure during subclinical viral myocarditis. Immunity. 2017 Nov 21;47(5):974-89.

[86] Bansal SS, Ismahil MA, Goel M, Patel B, Hamid T, Rokosh G, Prabhu SD. Activated T lymphocytes are essential drivers of pathological remodeling in ischemic heart failure. Circulation: Heart Failure. 2017 Mar;10(3):e003688.

[87] Borg N, Alter C, Görldt N, Jacoby C, Ding Z, Steckel B, Quast C, Bönner F, Friebe D, Temme S, Flögel U. CD73 on T cells orchestrates cardiac wound healing after myocardial infarction by purinergic metabolic reprogramming. Circulation. 2017 Jul 18;136(3):297-313.

[88] Cordero-Reyes AM, Youker KA, Trevino AR, Celis R, Hamilton DJ, Flores-Arredondo JH, Orrego CM, Bhimaraj A, Estep JD, Torre-Amione G. Full expression of cardiomyopathy is partly dependent on B-cells: a pathway that involves cytokine activation, immunoglobulin deposition, and activation of apoptosis. Journal of the American Heart Association. 2016 Jan 14;5(1):e002484.

[89] Frangou E, Vassilopoulos D, Boletis J, Boumpas DT. An emerging role of neutrophils and NETosis in chronic inflammation and fibrosis in systemic lupus erythematosus (SLE) and ANCA-associated vasculitides (AAV): implications for the pathogenesis and treatment. Autoimmunity reviews. 2019 Aug 1;18(8):751-60.

[90] Mutua V, Gershwin LJ. A Review of Neutrophil Extracellular Traps (NETs) in Disease: Potential Anti-NETs Therapeutics. Clinical Reviews in Allergy & Immunology. 2020 Aug 1:1-8.

[91] Sollberger G, Tilley DO, Zychlinsky A. Neutrophil extracellular traps: the biology of chromatin externalization. Developmental cell. 2018 Mar 12;44(5):542-53.

[92] Neubert E, Meyer D, Kruss S, Erpenbeck L. The power from within–understanding the driving forces of neutrophil extracellular trap formation. Journal of Cell Science. 2020 Mar 1;133(5).

[93] Németh T, Sperandio M, Mócsai A. Neutrophils as emerging therapeutic targets. Nature Reviews Drug Discovery. 2020 Jan 22:1-23.

[94] Becker RC, Owens AP, Sadayappan S. Tissue-level inflammation and ventricular remodeling in hypertrophic cardiomyopathy. Journal of thrombosis and thrombolysis. 2020 Feb 1;49(2):177-83.

[95] Fousert E, Toes R, Desai J. Neutrophil Extracellular Traps (NETs) Take the Central Stage in Driving Autoimmune Responses. Cells. 2020 Apr;9(4):915.

[96] Frangou E, Vassilopoulos D, Boletis J, Boumpas DT. An emerging role of neutrophils and NETosis in chronic inflammation and fibrosis in systemic lupus erythematosus (SLE) and ANCA-associated vasculitides (AAV): implications for the pathogenesis and treatment. Autoimmunity reviews. 2019 Aug 1;18(8):751-60.

[97] Stakos D, Skendros P, Konstantinides S, Ritis K. Traps N'Clots: NET-Mediated Thrombosis and Related Diseases. Thrombosis and haemostasis. 2020 Mar;120(03):373-83.

[98] Koitabashi N, Danner T, Zaiman AL, Pinto YM, Rowell J, Mankowski J, Zhang D, Nakamura T, Takimoto E, Kass DA. Pivotal role of cardiomyocyte TGF-β signaling in the murine pathological response to sustained pressure overload. The Journal of clinical investigation. 2011 Jun 1;121(6):2301-12.

[99] Ock S, Lee WS, Ahn J, Kim HM, Kang H, Kim HS, Jo D, Abel ED, Lee TJ, Kim J. Deletion of IGF-1 receptors in cardiomyocytes attenuates cardiac aging in male mice. Endocrinology. 2016 Jan 1;157(1):336-45.

[100] Adiarto S, Heiden S, Vignon-Zellweger N, Nakayama K, Yagi K, Yanagisawa M, Emoto N. ET-1 from endothelial cells is required for complete angiotensin II-induced cardiac fibrosis and hypertrophy. Life sciences. 2012 Oct 15;91(13-14):651-7.

[101] Aisagbonhi O, Rai M, Ryzhov S, Atria N, Feoktistov I, Hatzopoulos AK. Experimental myocardial infarction triggers canonical Wnt signaling and endothelial-to-mesenchymal transition. Disease models & mechanisms. 2011 Jul 1;4(4):469-83.

[102] Wei H, Bedja D, Koitabashi N, Xing D, Chen J, Fox-Talbot K, Rouf R, Chen S, Steenbergen C, Harmon JW, Dietz HC. Endothelial expression of hypoxia-inducible factor 1 protects the murine heart and aorta from pressure overload by suppression of TGF-β signaling. Proceedings of the National Academy of Sciences. 2012 Apr 3;109(14):E841-50

[103] Frangogiannis NG, Mendoza LH, Lewallen M, Michael LH, Smith CW, Entman ML. Induction and suppression of interferon-inducible protein (IP)-10 in reperfused myocardial infarcts may regulate angiogenesis. The FASEB Journal. 2001 Jun;15(8):1428-30.

[104] Bujak M, Dobaczewski M, Gonzalez-Quesada C, Xia Y, Leucker T, Zymek P, Veeranna V, Tager AM, Luster AD, Frangogiannis NG. Induction of the CXC chemokine interferon-γ-inducible protein 10 regulates the reparative response following myocardial infarction. Circulation research. 2009 Nov 6;105(10):973-83.

[105] Everett BM, MacFadyen JG, Thuren T, Libby P, Glynn RJ, Ridker PM. Inhibition of interleukin-1β and reduction in atherothrombotic cardiovascular Events in the CANTOS trial. Journal of the American College of Cardiology. 2020 Oct 6.

[106] Adamo L, Rocha-Resende C, Prabhu SD, Mann DL. Reappraising the role of inflammation in heart failure. Nature Reviews Cardiology. 2020 Jan 22:1-7.

[107] Okyere AD, Tilley DG. Leukocyte-Dependent Regulation of Cardiac Fibrosis. Frontiers in Physiology. 2020 Apr 8;11:301.

[108] Aghajanian H, Kimura T, Rurik JG, Hancock AS, Leibowitz MS, Li L, Scholler J, Monslow J, Lo A, Han W, Wang T. Targeting cardiac fibrosis with engineered T cells. Nature. 2019 Sep;573(7774):430-3.

[109] Vagnozzi RJ, Johansen AK, Molkentin JD. CARdiac Immunotherapy: T Cells Engineered to Treat the Fibrotic Heart. Molecular Therapy. 2019 Nov 6;27(11):1869-71.

[110] Nicolás-Ávila JA, Lechuga-Vieco AV, Esteban-Martínez L, Sánchez-Díaz M, Díaz-García E, Santiago DJ, Rubio-Ponce A, Li JL, Balachander A, Quintana JA, Martínez-de-Mena R. A network of macrophages supports mitochondrial homeostasis in the heart. Cell. 2020 Sep 15.

[111] Corden B, Lim WW, Song W, Chen X, Ko NS, Su L, Tee NG, Adami E, Schafer S, Cook SA. Therapeutic Targeting of Interleukin-11 Signalling Reduces Pressure Overload–Induced Cardiac Fibrosis in Mice. Journal of Cardiovascular Translational Research. 2020 Jun 26:1-7.

[112] Hao K, Lei W, Wu H, Wu J, Yang Z, Yan S, Lu XA, Li J, Xia X, Han X, Deng W. LncRNA-Safe contributes to cardiac fibrosis through Safe-Sfrp2-HuR complex in mouse myocardial infarction. Theranostics. 2019;9(24):7282

[113] Govindappa PK, Patil M, Garikipati VN, Verma SK, Saheera S, Narasimhan G, Zhu W, Kishore R, Zhang J, Krishnamurthy P. Targeting exosome-associated human antigen R attenuates fibrosis and inflammation in diabetic heart. The FASEB Journal. 2020 Feb;34(2):2238-51.

[114] Chothani S, Schäfer S, Adami E, Viswanathan S, Widjaja AA, Langley SR, Tan J, Wang M, Quaife NM, Jian Pua C, D'Agostino G. Widespread translational control of fibrosis in the human heart by RNA-binding proteins. Circulation. 2019 Sep 10;140(11):937-51.

[115] Cao Y, Liu C, Wang Q, Wang W, Tao E, Wan L. Pum2 mediates Sirt1 mRNA decay and exacerbates hypoxia/reoxygenation-induced cardiomyocyte apoptosis. Experimental Cell Research. 2020 May 8:112058.

[116] Ji X, Ding W, Xu T, Zheng X, Zhang J, Liu M, Liu G, Wang J. MicroRNA-31-5p attenuates doxorubicin-induced cardiotoxicity via quaking and circular RNA Pan3. Journal of Molecular and Cellular Cardiology. 2020 Mar 3.

[117] Yan F, Liu R, Zhuang X, Li R, Shi H, Gao X. Salidroside Attenuates Doxorubicin-Induced Cardiac Dysfunction Partially Through Activation of QKI/FoxO1 Pathway. Journal of Cardiovascular Translational Research. 2020 Jul 16:1-0.

[118] Xia W, Zou C, Chen H, Xie C, Hou M. Immune checkpoint inhibitor induces cardiac injury through polarizing macrophages via modulating microRNA-34a/Kruppel-like factor 4 signaling. Cell death & disease. 2020 Jul 24;11(7):1-5

[119] Gao C, Wang Y. mRNA Metabolism in Cardiac Development and Disease: Life After Transcription. Physiological Reviews. 2020 Apr 1;100(2):673-94.

[120] Rogers RG, Ciullo A, Marbán E, Ibrahim AG. Extracellular Vesicles as Therapeutic Agents for Cardiac Fibrosis. Frontiers in Physiology. 2020;11:479.

[121] Leonard RJ, Preston CC, Gucwa ME, Afeworki Y, Selya AS, Faustino RS. Protein subdomain enrichment of NUP155 variants identify a novel predicted pathogenic hotspot. Frontiers in cardiovascular medicine. 2020 Feb 7;7:8.

[122] Mancuso T, Barone A, Salatino A, Molinaro C, Marino F, Scalise M, Torella M, De Angelis A, Urbanek K, Torella D, Cianflone E. Unravelling the Biology of Adult Cardiac Stem Cell-Derived Exosomes to Foster Endogenous Cardiac Regeneration and Repair. International Journal of Molecular Sciences. 2020 Jan;21(10):3725.

[123] Hocine HR, Brunel S, Chen Q, Giustiniani J, San Roman MJ, Ferrat YJ, Palacios I, de la Rosa O, Lombardo E, Bensussan A, Charron D. Extracellular vesicles released by allogeneic human cardiac stem/progenitor cells as part of their therapeutic benefit. Stem cells translational medicine. 2019 Sep;8(9):911-24.

[124] Liang B, He X, Zhao YX, Zhang XX, Gu N. Advances in Exosomes Derived from Different Cell Sources and Cardiovascular Diseases. BioMed Research International. 2020 Jul 7;2020.

[125] Cai L, Chao G, Li W, Zhu J, Li F, Qi B, Wei Y, Chen S, Zhou G, Lu X, Xu J. Activated CD4+ T cells-derived exosomal miR-142-3p boosts post-ischemic ventricular remodeling by activating myofibroblast. Aging (Albany NY). 2020 Apr 30;12(8):7380.

[126] Shanmuganathan M, Vughs J, Noseda M, Emanueli C. Exosomes: basic biology and technological advancements suggesting their potential as ischemic heart disease therapeutics. Frontiers in Physiology. 2018 Nov 19;9:1159.

[127] Ranjan P, Kumari R, Verma SK. Cardiac Fibroblasts and Cardiac Fibrosis: Precise Role of Exosomes. Frontiers in Cell and Developmental Biology. 2019;7.

[128] Tikhomirov R, Donnell BR, Catapano F, Faggian G, Gorelik J, Martelli F, Emanueli C. Exosomes: from potential culprits to new therapeutic promise in the setting of cardiac fibrosis. Cells. 2020 Mar;9(3):592.

[129] Zhang Z, Wan J, Liu X, Zhang W. Strategies and technologies

for exploring long noncoding RNAs in heart failure. Biomedicine & Pharmacotherapy. 2020 Nov 1;131:110572.

[130] Wang Y, Liu B. Circular RNA in Diseased Heart. Cells. 2020 May;9(5):1240.

[131] Fioravanti A, Pirtoli L, Giordano A, Dotta F. Crosstalk between MicroRNA and Oxidative Stress in Physiology and Pathology.

[132] Chen C, Tang Y, Sun H, Lin X, Jiang B. The roles of long noncoding RNAs in myocardial pathophysiology. Bioscience Reports. 2019 Nov 29;39(11).

[133] Chen G, Huang S, Song F, Zhou Y, He X. Lnc-Ang362 is a pro-fibrotic long non-coding RNA promoting cardiac fibrosis after myocardial infarction by suppressing Smad7. Archives of Biochemistry and Biophysics. 2020 Mar 30:108354

[134] Liang H, Pan Z, Zhao X, Liu L, Sun J, Su X, Xu C, Zhou Y, Zhao D, Xu B, Li X. LncRNA PFL contributes to cardiac fibrosis by acting as a competing endogenous RNA of let-7d. Theranostics. 2018;8(4):1180

[135] Wawrzyniak O, Zarębska Ż, Kuczyński K, Gotz-Więckowska A, Rolle K. Protein-Related Circular RNAs in Human Pathologies. Cells. 2020 Aug;9(8):1841

[136] Lin R, Rahtu-Korpela L, Magga J, Ulvila J, Swan J, Kemppi A, Pakanen L, Porvari K, Huikuri H, Junttila J, Kerkelä R. miR-1468-3p promotes aging-related cardiac fibrosis. Molecular Therapy-Nucleic Acids. 2020 Apr 8.

[137] Li J, Cao LT, Liu HH, Yin XD, Wang J. Long non coding RNA H19: An emerging therapeutic target in fibrosing diseases. Autoimmunity. 2020 Jan 2;53(1):1-7.

[138] Wang X, Morelli MB, Matarese A, Sardu C, Santulli G. Cardiomyocyte-derived exosomal microRNA-92a mediates post-ischemic myofibroblast activation both in vitro and ex vivo. ESC heart failure. 2020 Feb;7(1):285-9

[139] Henning RJ. Cardiovascular exosomes and MicroRNAs in cardiovascular physiology and pathophysiology. Journal of cardiovascular translational research. 2020 Jun 25:1-8

[140] Ferrari S, Pesce M. Cell-Based Mechanosensation, Epigenetics, and Non-Coding RNAs in Progression of Cardiac Fibrosis. International Journal of Molecular Sciences. 2020 Jan;21(1):28.

[141] Morelli MB, Shu J, Sardu C, Matarese A, Santulli G. Cardiosomal microRNAs are essential in post-infarction myofibroblast phenoconversion. International journal of molecular sciences. 2020 Jan;21(1):201.

[142] Guo W, Zhu C, Yin Z, Wang Q, Sun M, Cao H, Greaser ML. Splicing factor RBM20 regulates transcriptional network of titin associated and calcium handling genes in the heart. International journal of biological sciences. 2018;14(4):369

[143] Picca A, Guerra F, Calvani R, Marini F, Biancolillo A, Landi G, Beli R, Landi F, Bernabei R, Bentivoglio AR, Monaco MR. Mitochondrial signatures in circulating extracellular vesicles of older adults with Parkinson's disease: Results from the EXosomes in PArkiNson's Disease (EXPAND) study. Journal of clinical medicine. 2020 Feb;9(2):504.

[144] Lin R, Rahtu-Korpela L, Magga J, Ulvila J, Swan J, Kemppi A, Pakanen L, Porvari K, Huikuri H, Junttila J, Kerkelä R. miR-1468-3p promotes aging-related cardiac fibrosis.

Molecular Therapy-Nucleic Acids. 2020 Apr 8.

[145] Zhou Y, Richards AM, Wang P. MicroRNA-221 is cardioprotective and anti-fibrotic in a rat model of myocardial infarction. Molecular Therapy-Nucleic Acids. 2019 Sep 6;17:185-97

[146] Li L, Bounds KR, Chatterjee P, Gupta S. Micro RNA-130a, a Potential Antifibrotic Target in Cardiac Fibrosis. Journal of the American Heart Association. 2017 Nov 7;6(11):e006763.

[147] Du W, Liang H, Gao X, Li X, Zhang Y, Pan Z, Li C, Wang Y, Liu Y, Yuan W, Ma N. MicroRNA-328, a potential anti-fibrotic target in cardiac interstitial fibrosis. Cellular Physiology and Biochemistry. 2016;39(3):827-36.

[148] Zhao X, Wang K, Liao Y, Zeng Q, Li Y, Hu F, Liu Y, Meng K, Qian C, Zhang Q, Guan H. MicroRNA-101a inhibits cardiac fibrosis induced by hypoxia via targeting TGFβRI on cardiac fibroblasts. Cellular Physiology and Biochemistry. 2015;35(1):213-26.

[149] Liu X, Xu Y, Deng Y, Li H. MicroRNA-223 regulates cardiac fibrosis after myocardial infarction by targeting RASA1. Cellular Physiology and Biochemistry. 2018;46(4):1439-54.

[150] Huang W, Feng Y, Liang J, Yu H, Wang C, Wang B, Wang M, Jiang L, Meng W, Cai W, Medvedovic M. Loss of microRNA-128 promotes cardiomyocyte proliferation and heart regeneration. Nature communications. 2018 Feb 16;9(1):1-6.

[151] Nishiga M, Horie T, Kuwabara Y, Nagao K, Baba O, Nakao T, Nishino T, Hakuno D, Nakashima Y, Nishi H, Nakazeki F. MicroRNA-33 controls adaptive fibrotic response in the remodeling heart by preserving lipid raft cholesterol. Circulation research. 2017 Mar 3;120(5):835-47.

[152] Yang HD, Nam SW. Pathogenic diversity of RNA variants and RNA variation-associated factors in cancer development. Experimental & Molecular Medicine. 2020 Apr 28:1-2.

[153] Fochi S, Lorenzi P, Galasso M, Stefani C, Trabetti E, Zipeto D, Romanelli MG. The Emerging Role of the RBM20 and PTBP1 Ribonucleoproteins in Heart Development and Cardiovascular Diseases. Genes. 2020 Apr;11(4):402

[154] Ranum PT, Goodwin AT, Yoshimura H, Kolbe DL, Walls WD, Koh JY, He DZ, Smith RJ. Insights into the biology of hearing and deafness revealed by single-cell RNA sequencing. Cell reports. 2019 Mar 12;26(11):3160-71.

[155] Zhang H, Zeitz MJ, Wang H, Niu B, Ge S, Li W, Cui J, Wang G, Qian G, Higgins MJ, Fan X. Long noncoding RNA-mediated intrachromosomal interactions promote imprinting at the Kcnq1 locus. Journal of Cell Biology. 2014 Jan 6;204(1):61-75.

[156] Wallace E, Howard L, Liu M, O'Brien T, Ward D, Shen S, Prendiville T. Long QT syndrome: genetics and future perspective. Pediatric cardiology. 2019 Aug 22:1-2.

[157] Lee Y, Rio DC. Mechanisms and regulation of alternative pre-mRNA splicing. Annual review of biochemistry. 2015 Jun 2;84:291-323.

[158] Obeng EA, Stewart C, Abdel-Wahab O. Altered RNA processing in cancer pathogenesis and therapy. Cancer discovery. 2019 Nov 1;9(11):1493-510.

[159] Shenasa H, Hertel KJ. Combinatorial regulation of alternative splicing. Biochimica et Biophysica Acta (BBA)-Gene Regulatory Mechanisms. 2019 Nov 1;1862(11-12):194392.

[160] Ule J, Blencowe BJ. Alternative splicing regulatory networks: functions, mechanisms, and evolution. Molecular cell. 2019 Oct 17;76(2):329-45

[161] Gerull B, Atherton J, Geupel A, Sasse-Klaassen S, Heuser A, Frenneaux M, McNabb M, Granzier H, Labeit S, Thierfelder L. Identification of a novel frameshift mutation in the giant muscle filament titin in a large Australian family with dilated cardiomyopathy. Journal of molecular medicine. 2006 Jun 1;84(6):478-83.

[162] Cheng H, Zheng M, Peter AK, Kimura K, Li X, Ouyang K, Shen T, Cui L, Frank D, Dalton ND, Gu Y. Selective deletion of long but not short Cypher isoforms leads to late-onset dilated cardiomyopathy. Human molecular genetics. 2011 May 1;20(9):1751-62

[163] Tang ZZ, Sharma S, Zheng S, Chawla G, Nikolic J, Black DL. Regulation of the mutually exclusive exons 8a and 8 in the CaV1. 2 calcium channel transcript by polypyrimidine tract-binding protein. Journal of Biological Chemistry. 2011 Mar 25;286(12):10007-16.

[164] Makarenko I, Opitz CA, Leake MC, Neagoe C, Kulke M, Gwathmey JK, Del Monte F, Hajjar RJ, Linke WA. Passive stiffness changes caused by upregulation of compliant titin isoforms in human dilated cardiomyopathy hearts. Circulation research. 2004 Oct 1;95(7):708-16.

[165] Lahmers S, Wu Y, Call DR, Labeit S, Granzier H. Developmental control of titin isoform expression and passive stiffness in fetal and neonatal myocardium. Circulation research. 2004 Mar 5;94(4):505-13.

[166] Opitz CA, Linke WA. Plasticity of cardiac titin/connectin in heart development. Journal of Muscle Research & Cell Motility. 2005 Dec 1;26(6-8):333-42.

[167] Warren CM, Krzesinski PR, Campbell KS, Moss RL, Greaser ML. Titin isoform changes in rat myocardium during development. Mechanisms of development. 2004 Nov 1;121(11):1301-12..

[168] Yamasaki R, Wu Y, McNabb M, Greaser M, Labeit S, Granzier H. Protein kinase A phosphorylates titin's cardiac-specific N2B domain and reduces passive tension in rat cardiac myocytes. Circulation research. 2002 Jun 14;90(11):1181-8.

[169] Cazorla O, Freiburg A, Helmes M, Centner T, McNabb M, Wu Y, Trombitas K, Labeit S, Granzier H. Differential expression of cardiac titin isoforms and modulation of cellular stiffness. Circulation research. 2000 Jan 7;86(1):59-67.

[170] Itoh-Satoh M, Hayashi T, Nishi H, Koga Y, Arimura T, Koyanagi T, Takahashi M, Hohda S, Ueda K, Nouchi T, Hiroe M. Titin mutations as the molecular basis for dilated cardiomyopathy. Biochemical and biophysical research communications. 2002 Feb 22;291(2):385-93.

[171] Gerull B, Gramlich M, Atherton J, McNabb M, Trombitás K, Sasse-Klaassen S, Seidman JG, Seidman C, Granzier H, Labeit S, Frenneaux M. Mutations of TTN, encoding the giant muscle filament titin, cause familial dilated cardiomyopathy. Nature genetics. 2002 Feb;30(2):201-4.

[172] Neagoe C, Kulke M, del Monte F, Gwathmey JK, de Tombe PP, Hajjar RJ, Linke WA. Titin isoform switch in ischemic human heart disease. Circulation. 2002 Sep 10;106(11):1333-41

[173] Williams L, Howell N, Pagano D, Andreka P, Vertesaljai M, Pecor T,

Frenneaux M, Granzier H. Titin isoform expression in aortic stenosis. Clinical Science. 2009 Sep 1;117(6):237-42.

[174] Chaturvedi RR, Herron T, Simmons R, Shore D, Kumar P, Sethia B, Chua F, Vassiliadis E, Kentish JC. Passive stiffness of myocardium from congenital heart disease and implications for diastole. Circulation. 2010 Mar 2;121(8):979.

[175] Wu Y, Peng J, Campbell KB, Labeit S, Granzier H. Hypothyroidism leads to increased collagen-based stiffness and re-expression of large cardiac titin isoforms with high compliance. Journal of molecular and cellular cardiology. 2007 Jan 1;42(1):186-95.

[176] Krüger M, Sachse C, Zimmermann WH, Eschenhagen T, Klede S, Linke WA. Thyroid hormone regulates developmental titin isoform transitions via the phosphatidylinositol-3-kinase/AKT pathway. Circulation research. 2008 Feb 29;102(4):439-47.

[177] Klein I, Ojamaa K. Thyroid hormone and the cardiovascular system. New England Journal of Medicine. 2001 Feb 15;344(7):501-9

[178] Krüger M, Babicz K, von Frieling-Salewsky M, Linke WA. Insulin signaling regulates cardiac titin properties in heart development and diabetic cardiomyopathy. Journal of molecular and cellular cardiology. 2010 May 1;48(5):910-6.

[179] Guo W, Schafer S, Greaser ML, Radke MH, Liss M, Govindarajan T, Maatz H, Schulz H, Li S, Parrish AM, Dauksaite V. RBM20, a gene for hereditary cardiomyopathy, regulates titin splicing. Nature medicine. 2012 May;18(5):766-73.

[180] Ma G, Samad I, Motz K, et al. Metabolic variations in normal and fibrotic human laryngotracheal-derived fibroblasts: a Warburg-like effect. Laryngoscope 2017;127:E107-13.

[181] Weeland CJ, van den Hoogenhof MM, Beqqali A, Creemers EE. Insights into alternative splicing of sarcomeric genes in the heart. Journal of molecular and cellular cardiology. 2015 Apr 1;81:107-13.

[182] Rexiati M, Sun M, Guo W. Muscle-specific Mis-splicing and heart disease exemplified by RBM20. Genes. 2018 Jan;9(1):18.

[183] Brauch KM, Karst ML, Herron KJ, De Andrade M, Pellikka PA, Rodeheffer RJ, Michels VV, Olson TM. Mutations in ribonucleic acid binding protein gene cause familial dilated cardiomyopathy. Journal of the American College of Cardiology. 2009 Sep 1;54(10):930-41

[184] Dauksaite V, Gotthardt M. Molecular basis of titin exon exclusion by RBM20 and the novel titin splice regulator PTB4. Nucleic acids research. 2018 Jun 1;46(10):5227-38.

[185] Watanabe T, Kimura A, Kuroyanagi H. Alternative splicing regulator RBM20 and cardiomyopathy. Frontiers in molecular biosciences. 2018 Nov 28;5:105.

[186] Morinaga A, Ito J, Niimi T, Maturana AD. RBM20 regulates CaV1. 2 surface expression by promoting exon 9* Inclusion of CACNA1C in neonatal rat cardiomyocytes. International journal of molecular sciences. 2019 Jan;20(22):5591.

[187] Cai H, Zhu C, Chen Z, Maimaiti R, Sun M, McCormick RJ, Lan X, Chen H, Guo W. Angiotensin ii influences pre-mRNA splicing regulation by enhancing RBM20 transcription through activation of the MAPK/ELK1 signaling pathway. International Journal of Molecular Sciences. 2019 Jan;20(20):5059.

[188] Briganti F, Sun H, Wei W, Wu J, Zhu C, Liss M, Karakikes I, Rego S, Cipriano A, Snyder M, Meder B. iPSC modeling of RBM20-deficient DCM identifies upregulation of RBM20 as a therapeutic strategy. Cell Reports. 2020 Sep 8;32(10):108117.

[189] Upadhyay SK, Mackereth CD. Structural basis of UCUU RNA motif recognition by splicing factor RBM20. Nucleic acids research. 2020 May 7;48(8):4538-50.

[190] Lennermann D, Backs J, van den Hoogenhof MM. New Insights in RBM20 Cardiomyopathy. Current Heart Failure Reports. 2020 Aug 13:1-3.

[191] Filippello A, Lorenzi P, Bergamo E, Romanelli MG. Identification of nuclear retention domains in the RBM20 protein. FEBS letters. 2013 Sep 17;587(18):2989-95.

[192] Murayama R, Kimura-Asami M, Togo-Ohno M, Yamasaki-Kato Y, Naruse TK, Yamamoto T, Hayashi T, Ai T, Spoonamore KG, Kovacs RJ, Vatta M. Phosphorylation of the RSRSP stretch is critical for splicing regulation by RNA-Binding Motif Protein 20 (RBM20) through nuclear localization. Scientific reports. 2018 Jun 12;8(1):1-4.

[193] Refaat MM, Lubitz SA, Makino S, Islam Z, Frangiskakis JM, Mehdi H, Gutmann R, Zhang ML, Bloom HL, MacRae CA, Dudley SC. Genetic variation in the alternative splicing regulator RBM20 is associated with dilated cardiomyopathy. Heart Rhythm. 2012 Mar 1;9(3):390-6.

[194] Li D, Morales A, Gonzalez-Quintana J, Norton N, Siegfried JD, Hofmeyer M, Hershberger RE. Identification of novel mutations in RBM20 in patients with dilated cardiomyopathy. Clinical and translational science. 2010 Jun;3(3):90-7.

[195] Wells QS, Becker JR, Su YR, Mosley JD, Weeke P, D'Aoust L, Ausborn NL, Ramirez AH, Pfotenhauer JP, Naftilan AJ, Markham L. Whole exome sequencing identifies a causal RBM20 mutation in a large pedigree with familial dilated cardiomyopathy. Circulation: Cardiovascular Genetics. 2013 Aug;6(4):317-26.

[196] Hey TM, Rasmussen TB, Madsen T, Aagaard MM, Harbo M, Mølgaard H, Møller JE, Eiskjær H, Mogensen J. Pathogenic RBM20-variants are associated with a severe disease expression in male patients with dilated cardiomyopathy. Circulation: Heart Failure. 2019 Mar;12(3):e005700.

[197] Robyns T, Willems R, Van Cleemput J, Jhangiani S, Muzny D, Gibbs R, Lupski JR, Breckpot J, Devriendt K, Corveleyn A. Whole exome sequencing in a large pedigree with DCM identifies a novel mutation in RBM20. Acta Cardiologica. 2019 Oct 3:1-6.

[198] Monaco I, Santacroce R, Casavecchia G, Correale M, Bottigliero D, Cordisco G, Leccese A, Di Biase M, Margaglione M, Brunetti ND. Double de novo mutations in dilated cardiomyopathy with cardiac arrest. Journal of Electrocardiology. 2019 Mar 1;53:40-3.

[199] Pantou MP, Gourzi P, Gkouziouta A, Tsiapras D, Zygouri C, Constantoulakis P, Adamopoulos S, Degiannis D. Phenotypic heterogeneity within members of a family carrying the same RBM20 mutation R634W. Cardiology. 2018;141(3):150-5.

[200] Sedaghat-Hamedani F, Haas J, Zhu F, Geier C, Kayvanpour E, Liss M, Lai A, Frese K, Pribe-Wolferts R, Amr A, Li DT. Clinical genetics and outcome of left ventricular non-compaction cardiomyopathy. European heart journal. 2017 Dec 7;38(46):3449-60.

[201] Long PA, Theis JL, Shih YH, Maleszewski JJ, Abell Aleff PC, Evans JM, Xu X, Olson TM. Recessive TAF1A mutations reveal ribosomopathy in siblings with end-stage pediatric dilated cardiomyopathy. Human molecular genetics. 2017 Aug 1;26(15):2874-81.

[202] Fedida J, Fressart V, Charron P, Surget E, Hery T, Richard P, Donal E, Keren B, Duthoit G, Hidden-Lucet F, Villard E. Contribution of exome sequencing for genetic diagnostic in arrhythmogenic right ventricular cardiomyopathy/dysplasia. PloS one. 2017 Aug 2;12(8):e0181840.

[203] Waldmüller S, Schroeder C, Sturm M, Scheffold T, Imbrich K, Junker S, Frische C, Hofbeck M, Bauer P, Bonin M, Gawaz M. Targeted 46-gene and clinical exome sequencing for mutations causing cardiomyopathies. Molecular and Cellular Probes. 2015 Oct 1;29(5):308-14.

[204] Liss M, Radke MH, Eckhard J, Neuenschwander M, Dauksaite V, von Kries JP, Gotthardt M. Drug discovery with an RBM20 dependent titin splice reporter identifies cardenolides as lead structures to improve cardiac filling. PloS one. 2018 Jun 11;13(6):e0198492.

[205] Maatz H, Jens M, Liss M, Schafer S, Heinig M, Kirchner M, Adami E, Rintisch C, Dauksaite V, Radke MH, Selbach M. RNA-binding protein RBM20 represses splicing to orchestrate cardiac pre-mRNA processing. The Journal of clinical investigation. 2014 Aug 1;124(8):3419-30

[206] Lorenzi P, Sangalli A, Fochi S, Dal Molin A, Malerba G, Zipeto D, Romanelli MG. RNA-binding proteins RBM20 and PTBP1 regulate the alternative splicing of FHOD3. The international journal of biochemistry & cell biology. 2019 Jan 1;106:74-83.

[207] Bertero A, Fields PA, Ramani V, Bonora G, Yardimci GG, Reinecke H, Pabon L, Noble WS, Shendure J, Murry CE. Dynamics of genome reorganization during human cardiogenesis reveal an RBM20-dependent splicing factory. Nature communications. 2019 Apr 4;10(1):1-9.

[208] Morinaga A, Ito J, Niimi T, Maturana AD. RBM20 regulates CaV1. 2 surface expression by promoting exon 9* Inclusion of CACNA1C in neonatal rat cardiomyocytes. International journal of molecular sciences. 2019 Jan;20(22):5591.

[209] Tharp C, Mestroni L, Taylor M. Modifications of Titin Contribute to the Progression of Cardiomyopathy and Represent a Therapeutic Target for Treatment of Heart Failure. Journal of Clinical Medicine. 2020 Sep;9(9):2770.

[210] Vikhorev PG, Smoktunowicz N, Munster AB, Kostin S, Montgiraud C, Messer AE, Toliat MR, Li A, Dos Remedios CG, Lal S, Blair CA. Abnormal contractility in human heart myofibrils from patients with dilated cardiomyopathy due to mutations in TTN and contractile protein genes. Scientific reports. 2017 Nov 1;7(1):1-1

[211] Hinson JT, Chopra A, Nafissi N, Polacheck WJ, Benson CC, Swist S, Gorham J, Yang L, Schafer S, Sheng CC, Haghighi A. Titin mutations in iPS cells define sarcomere insufficiency as a cause of dilated cardiomyopathy. Science. 2015 Aug 28;349(6251):982-6.

[212] Chopra A, Kutys ML, Zhang K, Polacheck WJ, Sheng CC, Luu RJ, Eyckmans J, Hinson JT, Seidman JG, Seidman CE, Chen CS. Force generation via β-cardiac myosin, titin, and α-actinin drives cardiac sarcomere assembly from cell-matrix adhesions. Developmental cell. 2018 Jan 8;44(1):87-96.

[213] Zhou Q, Kesteven S, Wu J, Aidery P, Gawaz M, Gramlich M, Feneley MP, Harvey RP. Pressure overload by transverse aortic constriction induces maladaptive hypertrophy in a titin-truncated mouse model. BioMed research international. 2015 Oct 4;2015.

[214] Ware JS, Cook SA. Role of titin in cardiomyopathy: from DNA variants to patient stratification. Nature Reviews Cardiology. 2018 Apr;15(4):241.

[215] Tabish AM, Azzimato V, Alexiadis A, Buyandelger B, Knöll R. Genetic epidemiology of titin-truncating variants in the etiology of dilated cardiomyopathy. Biophysical Reviews. 2017 Jun 1;9(3):207-23.

[216] Shibayama J, Yuzyuk TN, Cox J, Makaju A, Miller M, Lichter J, Li H, Leavy JD, Franklin S, Zaitsev AV. Metabolic remodeling in moderate synchronous versus dyssynchronous pacing-induced heart failure: integrated metabolomics and proteomics study. PLoS One. 2015 Mar 19;10(3):e0118974.

[217] Schafer S, De Marvao A, Adami E, Fiedler LR, Ng B, Khin E, Rackham OJ, Van Heesch S, Pua CJ, Kui M, Walsh R. Titin-truncating variants affect heart function in disease cohorts and the general population. Nature genetics. 2017 Jan;49(1):46-53

[218] Neishabouri SH, Hutson SM, Davoodi J. Chronic activation of mTOR complex 1 by branched chain amino acids and organ hypertrophy. Amino acids. 2015 Jun 1;47(6):1167-82.

[219] Methawasin M, Hutchinson KR, Lee EJ, Smith III JE, Saripalli C, Hidalgo CG, Ottenheijm CA, Granzier H. Experimentally increasing titin compliance in a novel mouse model attenuates the Frank-Starling mechanism but has a beneficial effect on diastole. Circulation. 2014 May 13;129(19):1924-36.

[220] Parikh VN, Caleshu C, Reuter C, Lazzeroni LC, Ingles J, Garcia J, McCaleb K, Adesiyun T, Sedaghat-Hamedani F, Kumar S, Graw S. Regional variation in RBM20 causes a highly penetrant arrhythmogenic cardiomyopathy. Circulation: Heart Failure. 2019 Mar;12(3):e005371.

[221] Wyles SP, Li X, Hrstka SC, Reyes S, Oommen S, Beraldi R, Edwards J, Terzic A, Olson TM, Nelson TJ. Modeling structural and functional deficiencies of RBM20 familial dilated cardiomyopathy using human induced pluripotent stem cells. Human molecular genetics. 2016 Jan 15;25(2):254-65.

[222] Streckfuss-Bömeke K, Tiburcy M, Fomin A, Luo X, Li W, Fischer C, Özcelik C, Perrot A, Sossalla S, Haas J, Vidal RO. Severe DCM phenotype of patient harboring RBM20 mutation S635A can be modeled by patient-specific induced pluripotent stem cell-derived cardiomyocytes. Journal of molecular and cellular cardiology. 2017 Dec 1;113:9-21.

[223] Motz K, Samad I, Yin LX, Murphy MK, Duvvuri M, Ding D, Hillel AT. Interferon-γ Treatment of Human Laryngotracheal Stenosis–Derived Fibroblasts. JAMA Otolaryngology–Head & Neck Surgery. 2017 Nov 1;143(11):1134-40.

[224] Hillel AT, Samad I, Ma G, et al. Dysregulated Macrophages Are Present in Bleomycin-Induced Murine Laryngotracheal Stenosis. Otolaryngol Head Neck Surg 2015;153:244-50

[225] Hillel AT, Ding D, Samad I, et al. T-Helper 2 Lymphocyte Immunophenotype Is Associated With Iatrogenic Laryngotracheal Stenosis. Laryngoscope 2019;129:177-86.

[226] Nesek-Adam V, Mršić V, Oberhofer D, Grizelj-Stojčić E, Košuta D, Rašić Ž. Post-intubation

long-segment tracheal stenosis of the posterior wall: a case report and review of the literature. Journal of Anesthesia. 2010;24(4):621-625.

[227] Hirshoren N, Eliashar R. Wound-healing modulation in upper airway stenosis-Myths and facts. Head & Neck. 2009;31(1):111-126.

[228] Antón-Pacheco J, Usategui A, Martínez I, García-Herrero C, Gamez A, Grau M et al. TGF-β antagonist attenuates fibrosis but not luminal narrowing in experimental tracheal stenosis. The Laryngoscope. 2016;127(3):561-567

[229] Loewen M, Walner D, Caldarelli D. Improved airway healing using transforming growth factor beta-3 In a rabbit model. Wound Repair and Regeneration. 2001;9(1):44-49.

[230] Yin LX, Motz KM, Samad I, et al. Fibroblasts in Hypoxic Conditions Mimic Laryngotracheal Stenosis. Otolaryngol Head Neck Surg 2017;156:886-892.

[231] Ghosh A, Malaisrie N, Leahy KP, et al. Cellular Adaptive Inflammation Mediates Airway Granulation in a Murine Model of Subglottic Stenosis. Otolaryngol Head Neck Surg 2011;144:927-33

[232] Wynn TA. Fibrotic disease and the TH1/TH2 paradigm. Nat Rev Immunol 2004;4:583-94.

[233] Haft S, Lee JY, Ghosh A, et al. Inflammatory Protein Expression in Human Subglottic Stenosis Tissue Mirrors That in a Murine Model. Ann Otol Rhinol Laryngol 2014;123:65-70

[234] Kendall RT, Feghali-Bostwick CA. Fibroblasts in fibrosis: novel roles and mediators. Front Pharmacol 2014;5:123

[235] Singh T, Sandulache V, Otteson T, Barsic M, Klein E, Dohar J et al. Subglottic Stenosis Examined as a Fibrotic Response to Airway Injury Characterized by Altered Mucosal Fibroblast Activity. Archives of Otolaryngology–Head & Neck Surgery. 2010;136(2):163

[236] Doolin E, Tsuno K, Strande L, Santos M. Pharmacologic Inhibition of Collagen in an Experimental Model of Subglottic Stenosis. Annals of Otology, Rhinology & Laryngology. 1998;107(4):275-279.

[237] Ma G, Samad I, Motz K, et al. Metabolic variations in normal and fibrotic human laryngotracheal-derived fibroblasts: a Warburg-like effect. Laryngoscope 2017;127:E107-13.

[238] Anis M, Zhao Z, Khurana J, Krynetskiy E, Soliman A. Translational genomics of acquired laryngotracheal stenosis. The Laryngoscope. 2014;124(5):E175-E179

[239] Boehler A, Chamberlain D, Kesten S, Slutsky AS, Liu M, Keshavjee S. Lymphocytic airway infiltration as a precursor to fibrous obliteration in a rat model of bronchiolitis obliterans. Transplantation. 1997;64(2):311-7.

[240] Sandulache VC, Chafin B, Li-Korotky HS, Otteson TD, Dohar JE, Hebda PA. Elucidating the Role of Interleukin 1ß and Prostaglandin E^2 in Upper Airway Mucosal Wound Healing. Acta Otolaryngol Head Neck Surg. 2007;133:365-74.

[241] Nicolli EA, Carey RM, Farquhar D, Haft S, Alfonso KP, Mirza N. Risk factors for adult acquired subglottic stenosis. J Laryngol Otol. 2017;131(3):264-7.

[242] Branski RC, Rosen CA, Verdolini K HP. Markers of wound healing in vocal fold secretions from patients with laryngeal pathology. Ann Otol Rhinol Laryngol 2004;113:23-9.

[243] Sharma M, Panda NK. Proteomic Profiling of Protease-Primed Virus-Permissive Caco-2 Cells Display Abortive-Interferon Pathway and Deregulated Thromboinflammatory SERPINS. Preprints. 2020 Jun 17.

[244] Zhao YG, Zhang H. Phase separation in membrane biology: the interplay between membrane-bound organelles and membraneless condensates. Developmental Cell. 2020 Jul 28.

[245] Dodson AE, Kennedy S. Phase separation in germ cells and development. Developmental Cell. 2020 Oct 1.

[246] Singh J. Phase Separation of RNA Helicase Triggers Stress-Responsive Translational Switch. Trends in Biochemical Sciences. 2020 Sep 1;45(9):726-8.

[247] Majumder S, Jain A. Osmotic Stress Triggers Phase Separation. Molecular Cell. 2020 Sep 17;79(6):876-7.

[248] Zbinden A, Pérez-Berlanga M, De Rossi P, Polymenidou M. Phase separation and neurodegenerative diseases: a disturbance in the Force. Developmental Cell. 2020 Oct 12;55(1):45-68.

[249] Jalihal AP, Pitchiaya S, Xiao L, Bawa P, Jiang X, Bedi K, Parolia A, Cieslik M, Ljungman M, Chinnaiyan AM, Walter NG. Multivalent proteins rapidly and reversibly phase-separate upon osmotic cell volume change. Molecular Cell. 2020 Sep 17;79(6):978-90.

[250] Szabo G, Momen-Heravi F. Extracellular Vesicles and Exosomes: Biology and Pathobiology. The Liver: Biology and Pathobiology. 2020 Feb 12:1022-7.

[251] Saha P, Sharma S, Mishra R, Guneshekaran M, Kaushal S, Davis ME, Vallabhajosyula P. Circulating Transplanted Progenitor Cell Specific Exosomes Predict Functional Recovery of Ischemic Myocardium. Circulation. 2018 Nov 6;138(Suppl_1):A16237-.

[252] Zhang ZC, Liu Y, Xiao LL, Li SF, Jiang JH, Zhao Y, Qian SW, Tang QQ, Li X. Upregulation of miR-125b by estrogen protects against non-alcoholic fatty liver in female mice. Journal of hepatology. 2015 Dec 1;63(6):1466-75.

Chapter 2

Extracellular Matrix in Cardiac Tissue Mechanics and Physiology: Role of Collagen Accumulation

Kristen LeBar and Zhijie Wang

Abstract

The extracellular matrix (ECM) forms a mesh surrounding tissue, made up of fibrous and non-fibrous proteins that contribute to the cellular function, mechanical properties of the tissue and physiological function of the organ. The cardiac ECM remodels in response to mechanical alterations (e.g., pressure overload, volume overload) or injuries (e.g., myocardial infarction, bacterial infection), which further leads to mechanical and functional changes of the heart. Collagen, the most prevalent ECM protein in the body, contributes significantly to the mechanical behavior of myocardium during disease progression. Alterations in collagen fiber morphology and alignment, isoform, and cross-linking occur during the progression of various cardiac diseases. Acute or compensatory remodeling of cardiac ECM maintains normal cardiac function. However, chronic or decompensatory remodeling eventually results in heart failure, and the exact mechanism of transition into maladaptation remains unclear. This review aims to summarize the primary role of collagen accumulation (fibrosis) in heart failure progression, with a focus on its effects on myocardial tissue mechanical properties and cellular and organ functions.

Keywords: collagen deposition, fibrosis, myocardial stiffening, left and right ventricle, mechanobiology

1. Introduction

The extracellular matrix (ECM) is a network of proteins, fibrous and non-fibrous, which form a supporting architecture for the cells in cardiac tissues. Cardiomyocytes, fibroblasts, vascular cells, and inflammatory cells that are responsible for the synthesis and degradation of ECM proteins exist in and around the cardiac ECM. A unique hallmark of the cardiovascular (CV) system is that the tissue is subject to dynamic mechanical load from the pulsatile blood pressure and flow. A perturbation of the mechanical load will be transduced to and sensed by the cells via the ECM to trigger acute or chronic remodeling of the tissue, resulting in structural and mechanical changes in the ECM. The altered ECM biomechanical properties further change the behavior of the cells within the tissue. These aspects are known as mechanobiology. This chapter will focus primarily on the changes of the ECM in various cardiac diseases, the alterations in the mechanical properties of the myocardium as a result of the ECM remodeling, and the impact of these

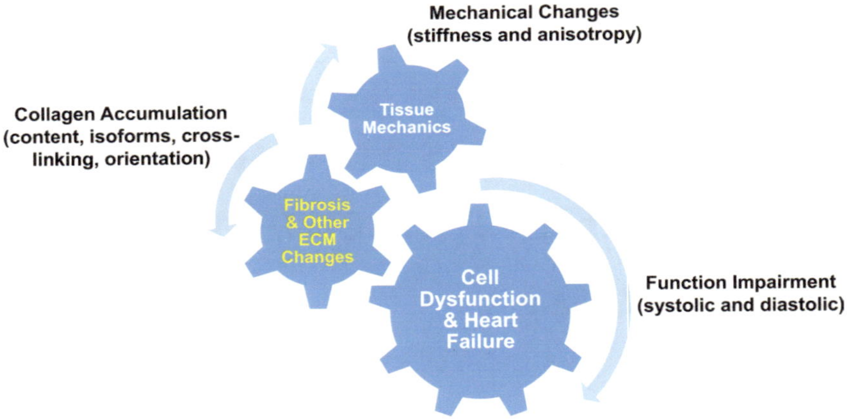

Figure 1.
A schematic plot showing the relations between cardiac ECM remodeling (fibrosis), tissue mechanics and ventricular function.

biomechanical factors on cellular and organ function in the progression of heart failure (**Figure 1**). Moreover, because collagen accumulation (fibrosis) is mostly investigated and contributes significantly to myocardial mechanical properties, we will limit our discussion on cardiac fibrosis as the main ECM remodeling event. Readers are referred to other extensive reviews that have summarized a broad category of ECM proteins during myocardial remodeling in different types of heart diseases [1, 2].

2. Extracellular matrix proteins in cardiac tissues

2.1 Overview of extracellular matrix proteins in biological tissues

Extracellular matrix (ECM) proteins found in biological tissue can serve structural or non-structural roles, depending on the location and composition of the protein. Glycoproteins primarily consist of structural ECM proteins, which include fibrillar collagen, elastin, fibronectin, and laminin [1]. These are often the main determinants of tissue's passive mechanical properties. The non-structural ECM proteins are primarily categorized as proteoglycans, which are further distinguished as four subgroups—hyalectans, cell surface proteoglycans, basement membrane proteoglycans, and small leucine-rich proteoglycans (SLRP's) [2]. These proteins play a key role in cell–cell or Cell—matrix interactions, interaction with growth factors, as well as binding to cell receptors to regulate cellular function [2].

Among the structural ECM proteins, collagen and elastin are mostly investigated in cardiovascular tissues. Collagen is the most abundant protein in the body and formed from the basic unit of tropocollagen [3]. Tropocollagens, made up of polypeptide chains, spontaneously twist together to form a triple helix structure, which form a newly synthesized procollagen fibril. Mature collagen fibers can then form via cross-linking of procollagen fibrils into bundles or sheets, conferring versatile mechanical properties in various tissues [3]. There are over 25 collagen types to date [2], each with different physical and mechanical properties [4]. For instance, type I collagen is a subtype that exhibits stiffer mechanical property and higher tensile strength compared to type III collagen. Collagen metabolism is maintained by the synthesis and degradation balance, and different tissues have different turnover rates [4, 5]. Higher turnover of collagen types I and III is observed in diseased tissues, and is also linked to the pathological state of the tissue such as inflammation and aging [4, 5]. Overall, fibrillar collagen contributes to CV tissue's nonlinear elastic behavior at high strains due to the increased recruitment of fibers.

Elastin is another primary structural protein, contributing to the mechanical property of the CV tissue at low strains [6]. It is formed as sheets comprised of the base unit tropoelastin, which has the ability to stretch and recoil [6]. Compared to collagen fibers which have an average Young's Modulus of 250–400 MPa [7], elastin fibers are more compliant with an average Young's Modulus of 1 MPa [8]. Elastin is a highly stable protein that has very low turnover rate – once formed, it lasts almost for the entire lifespan [9]. Increased elastin degradation or damage is a sign of pathological remodeling of the tissue and is associated with aging and CV diseases [9].

2.2 Primary ECM proteins in the heart

In the myocardial tissue, the main ECM protein is collagen. Collagen is diffusively spread over in the myocardium—interstitial and perivascular collagen fibers have been revealed by histology. Collagen types I and III are the most prominent collagen types in the myocardium [10], making up 85% and 6–11% of the total collagen, respectively [11]. Other isoforms of collagen are also reported. Bashey et al. examined murine, canine, and nonhuman primate healthy hearts and found that type V collagen comprises 2 ~ 3% and type VI collagen comprises ~5% of the total collagen in the myocardium [11]. While the fractional content of type IV collagen was not evident, it appeared most prominent in the basement membrane and in the media [11]. A detailed summary of all fibrillar and non-fibrillar collagen and their roles in cardiac diseases can be found in a recent review [12].

Elastin content in the ventricle is not detailed in the same manner as collagen. The distribution of elastin is mainly limited to the epicardium, the outer layer of the ventricular wall [13, 14]. Biochemical measurements showed that elastin content is about one tenth of the collagen content in healthy heart. Furthermore, the left ventricle (LV) tended to have more elastin and collagen content (in μg/mg) than the right ventricle (RV) [15].

In heart valves, elastin is more abundant than in the myocardium. It is located primarily in the inflow layer and sparsely distributed in the outflow and central layers, comprising approximately 10% of total proteins in the tissue [16, 17]. Collagen type I is found primarily in the valve leaflets and the valve outflow layer, whereas collagen type III is distributed throughout the entire valve structure [16]. Collagen comprises approximately 60% of ECM proteins in human heart valves [17].

2.3 Cells and molecules responsible for ECM synthesis and degradation

The cellular components responsible for the cardiac collagen synthesis include interstitial fibroblasts (in healthy hearts), transdifferentiated myofibroblasts and inflammatory cells (in diseased hearts), cardiomyocytes, and adventitial fibroblasts and smooth muscle cells (SMCs) in the blood vessels [18–20]. Several biological mediators such as pro-inflammatory cytokines, growth factors and hormones are also identified to participate in collagen synthesis, which are reviewed previously [4, 21, 22]. Among them, fibroblasts and transforming growth factor beta (TGF-β) are two main contributing factors. In cardiac remodeling, fibroblasts migrate to the injured region or to the area where ECM proteins are over-degraded and secrete new ECM proteins—primarily collagen types I and III [18]. Moreover, a "new" phenotype of cells myofibroblasts, emerge in injured myocardium, which is a key step to strengthen cardiac fibrosis in both infarcted and hypertensive myocardium.

MMPs	Substrate
MMP-1	Collagens I, II, III, VII, X
MMP-8	Collagens I, II, III, V, VII, X
MMP-13	Collagens I, II, III, IV, fibronectin, laminin
MMP-2	Collagens, I, IV, V, VII, X, XI, fibronectin, laminin elastin
MMP-9	Collagens III, IV, V, VII, X, elastin
MMP-3	Collagens III, IV, V, IX, X, fibronectin, laminin
MMP-10	Collagens III, IV, V, IX, laminin, fibronectin
MMP-11	Collagen IV, fibronectin, laminin
MMP-14	Collagens I, II, III, fibronectin, laminin

Table 1.
Matrix Metalloproteinases (MMPs) and their substrates (adapted from [24]).

TGF-β is involved in signaling pathways of various cells (fibroblasts, cardiomyocytes, immune and vascular cells) to initiate fibrogenic action [19]. For a thorough review of current understanding of cardiac fibrosis in ischemic and non-ischemic heart diseases, the reader is referred to these references [12, 21].

Degradation of the ECM proteins is necessary for the turnover as well as normal protein function. Matrix metalloproteinases (MMPs) are the most significant molecules that contribute to this degradation and these enzymes are key in the CV tissue remodeling [10]. MMPs in the heart are primarily expressed by the fibroblasts and cardiomyocytes [20, 23]. **Table 1** details various types of MMPs and the ECM proteins that they target. Insoluble fibrillar collagen such as collagen type I and III or more cross-linked collagen is difficult to be enzymatically degraded [11]. To prevent excessive degradation of ECM, tissue inhibitors of metalloproteinases (TIMPs) are secreted to bind to MMPs and limit the role of activated MMPs. Therefore, the overall balance between the activated MMPs and TIMPs determines the ECM remodeling and turnover rate.

3. Measurement of ECM proteins

There is increasing agreement that the ECM is not a passive biological component but actively interferes with cellular and organ function in the dynamic remodeling process. Thus, the measurement of ECM proteins is critical to study their roles in tissue biomechanics and remodeling in various diseases. The existing measurement methods can be classified into these categories: medical imaging techniques, optical imaging techniques, biochemical and biological methods.

3.1 Medical imaging techniques

Medical imaging techniques are generally noninvasive because they can be performed on a live subject with negligible risks; they are sometimes referred to as organ-scale imaging [25]. Primary medical imaging techniques include Magnetic Resonance Imaging (MRI), ultrasound and nuclear imaging. These techniques could measure bulk quantities of the materials—concentration, volume fraction and distribution of proteins.

MRI is the most common imaging method for collagen detection because of its higher sensitivity to specific molecular probes to target specific ECM proteins [26].

For example, the feasibility of detection of collagen (predominantly collagen type I) using a gadolinium-containing molecular contrast agent (delayed or late gadolinium enhancement) has been demonstrated in preclinical animal studies [27, 28]. The technique is valuable in the detection of fibrosis in ventricles [29, 30]. Recently, T1 mapping technique has emerged as a more useful technique for diffuse interstitial fibrosis measurement [31, 32]. In addition, a particular mode of MRI, tagging and feature tracking, enables clinicians to measure tissue strain from which collagen measurements can be indirectly deduced [33]. Elastin content can be quantified directly by molecular MRI as well [34].

Ultrasound technique is another imaging method to quantify collagen or elastin. Strain elastography measures elasticity of the myocardium, from which properties of the structure and content of collagen and elastin are indirectly derived [35, 36]. This method is sensitive to the fiber angle and density, both of which give light to the health status of the tissue [35]. Using the known mechanical properties of elastin and collagen (i.e., Young's Modulus), one can distinguish the relative contributions of ECM proteins to the tissue. Finally, cardiac nuclear imaging, including single-photon emission computed tomography (SPECT) and positron-emission tomography (PET), has been used to quantify the ECM content [25]. Radioactive molecular probes are used in these techniques. The common targets include collagen type I, II and IV, as well as MMPs [37, 38].

3.2 Optical imaging techniques

Optical imaging techniques are often used to provide the images of collagen fibers including their content and fiber orientation in intact, fresh tissues using physical properties of photons [39]. Two primary optical imaging techniques are the Second-harmonic Generation (SHG) and Small Angle Light Scattering (SALS). SHG imaging is a form of nonlinear optical microscopy. It can be applied to fresh or fixed tissues at varying depths (optical section), thus revealing the 2D or 3D structure of the collagen fibers [40]. This technique is advantageous due to its high resolution and specificity for the microstructure of collagen fibers [41]. However, it only allows the imaging of samples to a certain distance (depth), and deep tissue imaging systems are currently under development to enable larger length of penetration [42]. SALS is another method to measure fiber orientation when a polarized light beam is passed through a specimen—biological or nonbiological [43, 44]. The distribution of the scattered light is used to identify fiber orientation and alignment [43]. The gross/average collagen fiber orientation can be obtained in tissues with a thickness of at least 500 microns [45]. The advantage of this technique is the capability to measure in thicker tissues than SHG; but the disadvantage is that only an average of planar (2D) fiber orientation is derived, and the information along the depth direction is unavailable.

3.3 Biochemical and biological measurements

Direct measurements of collagen can be obtained using a long-established biochemical measurement – hydroxyproline assay, which quantifies the hydroxyproline content in digested samples. Hydroxyproline is a main molecular component of collagen and its amount can indirectly reflect the collagen content or is converted to collagen content with assumed collagen to hydroxyproline ratio [46]. Different methods have been established to measure hydroxyproline including colorimetric methods, high-performance liquid chromatography (HPLC) and liquid chromatography-mass spectrometry (LC–MS) [47].

In addition, tissue staining protocols—histology methods—are convenient to examine the ECM protein content and structure without losing the local distribution information. The common staining methods for collagen are Masson's Trichrome stains [35, 48] and Picrosirius Red stains [49]. Because mature collagen is birefringent and the Picrosirius Red stain can enhance the birefringency of collagen, collagen fibers can be visualized better and in more details under polarized light. It also enables a quick examination of types I and III collagen in CV tissues due to the different fiber thickness [50, 51]. Alternatively, the elastin is often examined by the Verhoeff-Van Gieson (VVG) stain [52] and the glycosaminoglycans (GAGs) and proteoglycans are examined by the Alcian blue stain [53]. Moreover, immunohistochemistry (IHC) employs the use of antibodies to quantify the specific protein of interest [10, 11].

Non-microscopic measurement methods include enzyme-linked immunosorbent assay (ELISA) and other standard biological methods. ELISA has been applied to detect collagen types I and III, as well as elastin and cross-linking of collagen [54, 55]. Finally, like all other proteins, the ECM proteins can be quantified by the protein expression using Western blot (immunoblotting) or by the mRNA or DNA expression using qPCR (quantitative Polymerase Chain Reaction) [6, 56].

4. Alterations of biomechanical properties in cardiac disease progression

Heart failure (HF) affects approximately 6.2 million adult Americans [57]. The main causes of HF are myocardial infarction, pressure-overload (hypertension), volume-overload, arrhythmia, valve stenosis or regurgitation, etc. Ventricular dysfunction is the most common type of HF including left-sided HF with preserved ejection fraction (HFpEF) and reduced ejection fraction (HFrEF), as well as right-sided HF secondary to pulmonary hypertension and congenital heart disease [58–60]. The malfunction of the myocardium can occur in the LV, RV, or both ventricles (biventricular HF).

As shown in the overall scheme (**Figure 1**), the cardiac ECM remodeling is an interactive, dynamic procedure that brings the cellular function, tissue mechanical behavior and organ function into one scenario. The ECM remodeling leads to both biological (structural) and mechanical (functional) changes of the tissue, which in turn regulates cell behaviors and alters the hemodynamics and cardiac performance. It is accepted that the remodeling often starts with an attempt to maintain the homeostasis environment of the cells and organ. This is referred to as adaptive remodeling (compensation). However, when the remodeling cannot achieve a 'stable' status of the new homeostasis but continues to deteriorate, impairments of the cells and organ will occur. This is known as adverse or mal-adaptive remodeling (decompensation). The mechanism of transition from compensation to decompensation remains a key knowledge gap. In almost all types of HF, cardiac fibrosis plays an important role in the pathogenesis; but its effects on mechanical changes of the myocardium and the physiological function are less noted. Hence, our discussion here will focus on the maintenance of the biomechanical homeostasis in common types of HF involving both LV and RV.

4.1 Biomechanical adaptations of hypertensive (pressure overload) myocardium

4.1.1 Fibrotic changes in hypertensive LV and RV

Hypertension is defined as chronically elevated blood pressure in the systemic circulation, with a systolic blood pressure greater than 120 mmHg, and/or a diastolic blood pressure greater than 80 mmHg [61]. It is one of the most prevalent pathologies in the United States, affecting approximately one in three adults. Hypertensive heart disease (HHD) accounts for approximately a quarter of all causes of heart failure [61]. In response to the pressure overload, cardiac hypertrophy and fibrosis are the most prominent events observed. The increased interstitial and perivascular collagen is originated from several cell types including cardiac fibroblasts, activated macrophages [62], cells derived from EndMT (or epithelial-to-mesenchymal transition (EMT)) or myofibroblasts, which are transdifferentiated from EndMT and EMT [21, 63].

In preclinical animal models, HHD can be induced by aortic banding [64, 65], genetic alteration (e.g. spontaneous hypertensive rates (SHR)) [66], or other methods that overlap with the models of systemic hypertension (e.g. high-salt diet or angiotensin II infusion) [63]. From both clinical and animal studies, various fibrillar collagen types such as I, III, IV, and V were reported to increase in the hypertensive LV [12, 64, 65, 67, 68]. The regulation of collagen turnover is dynamic. For instance, collagen types I and III increased within a day after hypertension was induced [64, 65]. Collagen types I and III reached a peak content (fivefold and 1.7-fold, respectively) at day 3, then decreased at day 7 and plateaued four weeks after the aortic banding. In contrast, collagen type IV reached its peak one day after the banding, and then began to decline at day 3 and plateaued at day 7 [65]. A similar trend of an initial elevation followed by a decline of collagen types I, III and IV has been reported by Jalil et al. in a rat aortic banding model [64]. The time-dependent change has been suspected to be associated with the transition from compensative hypertrophy to decompensation. The overall ECM degradation in later stages of HHD is speculated to disrupt the mechanical support and electrical conduction for myocardial contraction and promote cardiomyocyte apoptosis, which results in impairments of cardiomyocyte contractility and organ failure [12, 69].

There is no consensus on whether type I or type III collagen plays a more significant role in the LV hypertrophy. Some studies have indicated a predominance of type I collagen accumulation [70, 71], but others reported equivalent elevations of type I and III [72]. In a patient study that examined the collagen type I and III mRNAs expression in dilated cardiomyopathy, it was shown that the collagen type III/I ratio was higher in dilated cardiomyopathy patients (idiopathic, hypertensive and alcoholic) than the healthy controls [73]. This suggests an important role of collagen type III in the hypertensive LV fibrosis. The metabolism of the two collagen isoforms may not be independent since it has been shown that type III collagen is crucial for collagen I fibrillogenesis during the normal development of cardiovascular system [74]. Therefore, a further understanding of the role of collagen isoforms in fibrosis is needed.

In the RV, similar fibrotic remodeling occurs in response to pulmonary hypertension (PH), which is defined as a mean pulmonary arterial pressure (mPAP) exceeding 20 mmHg [75]. Increases in total collagen content or collagen synthesis were consistently reported in hypertensive RVs, from clinical to preclinical studies, from large to small animals, as well as from early to late stage of RV failure [50, 76–78]. The types of collagen upregulated in the RV has been investigated but inconsistent findings are reported. In a mouse model of PH (Sugen+hypoxia), the ratio of collagen type I/III was increased in the diseased RV and it was mainly attributed to the increase in type I collagen [78]. But in a recent ovine model of PH, type III collagen rather than type I collagen was found to be increased more significantly in the RV [51]. While both studies used Picrosirius Red stained histology samples to quantify collagen isoforms in the RV, these results need to be confirmed by other quantitative methods in future studies.

Finally, not only is there a change of the collagen content, but the morphology, cross-linking, and alignment are altered in the remodeling process. First, in the hypertensive LV, collagen type I became thicker and denser, creating a tighter mesh of fibers [64]. Second, the increase in collagen content is accompanied with an increase in cross-linking as this is part of the maturation of new collagen. Cross-linking is an enzymatic driven event and two types of enzymes—the LOX (lysyl oxidase) family and TG (tissue transglutaminase)—have been found to be upregulated in hypertensive myocardium [79, 80]. It has been shown that the collagen cross-linking, not content, was associated with the LV chamber stiffness and filling pressure [80, 81]. Collagen cross-linking was also associated with a higher incidence of hospitalization in HHD patients [82]. On the other hand, the cardiac remodeling requires degradation of insoluble collagens to enable rearrangement of cells and matrix proteins, and reduction in collagen cross-linking was reported as well [70]. Therefore, the role of cross-linking has not been fully understood in HHD. Third, collagen fibers became more aligned in the hypertensive LV [65]. Similarly, enhanced fiber alignment has been reported in the RV. In rat PH RVs, the myo-fibers and collagen fibers were more strongly aligned in the longitudinal (apex-to-outflow) direction so that the tissue became more anisotropic in mechanical behavior [83].

4.1.2 Mechanical changes in hypertensive LV and RV

Myocardial stiffening is widely evident in HHD patients, particularly revealed by the increase in diastolic (passive) stiffness of the LV [65, 67]. The reduced mechanical strain is a surrogate of myocardial stiffening and became noticeable in hypertensive LVs even when the contractility was preserved [32, 84]. This indicates that the stiffening occurs in the early stage of HHD. Increased myocardial stiffness is thought to predominantly contribute to the diastolic dysfunction, which is evident by increases in isovolumetric relaxation time (IVRT), and end-diastolic volume or diameter (EDV or EDD) [85, 86]. Moreover, the persistent diastolic dysfunction with a dilatation/thinning of the myocardium is associated with impaired contractile (systolic) function, which was revealed by changes in dP/dt_{max}, dP/dt_{min}, end-systolic pressure-volume relations (ESPVR), fractional shortening (FS), ejection fraction (EF), stroke volume (SV), or cardiac output (CO) [87]. The mechanical changes are of high clinical relevance as the myocardial stiffness is significantly greater in the high-risk patients than in healthy controls, which may indicate a transition to heart failure with preserved ejection fraction (HFpEF) [88]. However, what initiates the transition from adaptive to maladaptive remodeling remains unclear.

Like the LV, the RV myocardium stiffens under chronic pressure overload [83, 89, 90]. The clinical evidence of RV stiffening in PH patients has been reported via the myocardial strain or strain rate measurements [91, 92]. Compared with the LV, the RV passive mechanics seem to play a more important role in physiological function, which is supported by the findings that the RV elasticity is more closely associated with the severity of RV failure and is better related to prognosis than the RV systolic function [93–95]. The RV stiffening is also evident from *ex vivo* measurements of RV mechanical properties from a number of preclinical animal studies. A significant increase in RV stiffness was noted in the PH group in both longitudinal and circumferential directions, with or without cardiomyocytes [96, 97]. But some reported a greater change in stiffness in the longitudinal direction [90], whereas others reported a greater change in the circumferential direction [89]. Therefore, the characterization of anisotropic mechanical changes of hypertensive RV needs to be further elucidated.

The potential clinical significance of RV stiffening has been explored in a few studies. Jang et al. found that in PH rats, the RV longitudinal elastic modulus (EM) derived at low strains was correlated with RV diastolic function (end-diastolic elastance). This is the first report on the linkage of RV tissue mechanics and in vivo hemodynamics [90]. Recently, from an ovine model of PH, the RV longitudinal stiffness was significantly increased and correlated with the long-axis end-diastolic or end-systolic diameter or area. Moreover, in the longitudinal (apex-to-outflow tract) direction, there were trends of correlations between the low-strain EM and the acceleration time, as well as between the high-strain EM and the deceleration time. These findings indicate the critical role of the RV passive mechanical properties in the organ function [98]. Nevertheless, the question remains as to how exactly the mechanical changes affect the transition from adaptive to maladaptive remodeling.

4.1.3 Different roles of fibrosis in hypertensive LV and RV?

Finally, the role of RV fibrosis in PH development may be different than the fibrosis in the hypertensive LV. For instance, RV fibrosis occurs early with the pressure overload and no report of collagen degradation has been noted in failing RVs. In contrast, collagen degradation has been documented in the late stage of LV failure (see **Figure 3** below). Second, the RV fibrosis measured by T1 mapping was correlated with pulmonary arterial stiffness and RV RAC (relative area change), but not correlated with pulmonary pressure, RV mass or ejection fraction in PH patients [99]. This indicates that the RV fibrosis may be an early marker of maladaptive RV remodeling before the deterioration in the functional metric [99]. Such prognostic role has not been reported in LV fibrosis. Third, different outcomes of anti-fibrotic treatment were found between LV and RV. In the LV, treatments that induced reduction of fibrosis had led to regression of chamber stiffness and function improvement [71, 100–103]. The beneficial outcomes of anti-fibrotic treatment are convincing and recently reviewed [104]. But interestingly, interruptions of collagen accumulation in RV had led to various consequences. The restriction of collagen accumulation using a transgenic mouse model resulted in limited RV hypertrophy and preserved RF function in PH development, indicating a protective role of anti-fibrosis therapy for the RV [78]. Other drug studies that demonstrated reduced RV fibrosis and improved RV function are briefly summarized by Bogaard et al. [105]. But recently, an anti-fibrotic intervention via suppressed Gelectin-3 expression (knock-out mice) or pharmaceutical inhibitors was insufficient to improve RV function in PH mice, despite an improvement in RV fibrosis [106]. These results raise more questions about the role of RV fibrosis in its function. Therefore, whether and how the fibrotic event precedes the functional decline in the RV may be organ specific and remains to be elucidated. Other different responses of the LV and RV to pressure overload are listed in **Table 2**.

4.2 Biomechanical adaptations in volume overload myocardium

Volume overload is initially learned as physiological responses of myocardium because of the reversible myocardial remodeling observed in athletes and women in pregnancy. But pathological responses are also found in patients with heart valve disease (regurgitation) and congenital heart disease, which alternatively lead to heart failure. As a result, different views form regarding whether the remodeling from volume overload is adaptive and irreversible [110, 111]. In contrast to pressure overload, volume overload is often treated as another type of mechanical 'insult'

	RV	LV
Fibrosis	Persistent in all stages	Mostly in early stage; degradation occurs in late stage
Cardiomyocytes	↓ α-MHC and ↑β-MHC (fetal gene expression); no atrial natriuretic peptide (ANP); inefficient energy metabolism	↑β-MHC only; ↑ ANP; Improved energy metabolism
Collagen isoforms	Type I and III	Type I, III, V, IV, VIII, and more
Collagen fiber	More aligned fibers into apex-to-outflow tract direction	Thicker, denser collagen type I, more cross-linking & aligned fibers
Dilatation	Occurs acutely in early stage	Occurs in late stage
Capillary rarefaction? (↓ blood perfusion)	Yes	No
Inflammatory response involved?	Yes (macrophages, TNF-α, IL-1β)	Yes (T lymphocytes, TNF-α, IL-1β)
Response to anti-fibrosis treatment	Discrepant findings on function improvement	Mostly effective

Table 2.
Different responses of the LV and RV to pressure overload. MHC: myosin heavy chain; ANP: atrial natriuretic peptide. Adapted from [107–109].

in which the tissue is stretched beyond its normal state during diastole [112–114]. Under this type of mechanical load, collagen loss and chamber dilatation occurred since the early stage and these changes persisted, leading to an overall decrease in cardiac ECM [112–115]. Such remodeling was attributed to an increase in MMP activity [113]. But elastin showed biphasic changes in the progression of ECM remodeling. Ruzicka et al. reported an initial (within one week of induced volume overload) increase in elastin concentration but then a decrease of elastin concentration below control levels 10 weeks after induced volume overload [114]. The reduction of ECM turnover in volume overloaded LV was related to a phenotype change of fibroblasts into hypofibrotic type [116] and increased MMP expressions from macrophages or mast cells [117–119]. An increased collagen III/I ratio was also noted in the compensated stage of the remodeling [114]. The elevated ECM degradation in volume overloaded LV seems to share common pathways as seen in late stage of HHD—to enable ventricle dilatation and thinning and weaken the cell-matrix connections, which impairs contractile function.

The RV did not undergo initial remodeling to the same extent of the LV. It is well known that the RV is better in adaptation of the volume overload whereas the LV is better in adaptation of the pressure overload [107, 120]. The chamber difference also lies in the fact that LV alterations are more widely reported on than those of the RV and that the RV has exhibited milder remodeling than the LV [121]. The less pronounced remodeling including the ECM alteration of the RV may be explained by the different origins and contractile behaviors of cardiomyocytes [120, 122] and, thus, different responses to the mechanical stimulations.

Like in pressure overloaded HF, myocardial stiffening also arises in response to volume overload and eventually leads to heart failure [123, 124]. For instance, Emery et al. reported a 10-fold increase in LV mid-wall stiffness along the fiber direction and the cross-fiber direction six weeks after volume overload induction [125]. The underlying cause of the myocardial stiffening is investigated by collagen measurement. Interestingly, despite the decrease in the relative collagen content, there was an upregulation of collagen cross-linking [115]. A similar finding was identified using hydroxylysylpyridinoline (HP) assay in minipig LVs [126]. Therefore, despite a decrease in total collagen content, the tissue stiffening occurs due to elevated cross-linking in these ventricles [115, 125, 126].

Myocardial compliance, however, showed different trends of changes than the intrinsic (material) mechanical property of the myocardium. This is because the chamber compliance is a measurement of overall stiffness that incorporates changes in the intrinsic mechanical property, the geometry of the wall (dilatation and thinning) and the contractility of the heart (ventricular pressure). In the acute stage, the LV wall dilated and the compliance increased 2 days after induction of volume overload [112]. Similar findings are reported in the compensated stage. A significant decrease in the end-diastolic pressure-volume relationship (EDPVR) and a right shift of the pressure-volume loop were seen in the group with 8-week volume overload [127]. This indicated increased compliance and chamber dilatation. But at 15–21 weeks of volume overload, there was no significant change in the EDPVR compared to that at week 8 [127]. Thus, the maintained compliance may be a combined results of increased intrinsic stiffness and decreased myocardial thickness.

4.3 Different myocardial remodeling induced by pressure and volume overload

Both pressure overload and volume overload are categorized as the hemodynamic (mechanical) 'insult' of the myocardium. The pressure overload is considered as an afterload increase whereas the volume overload is considered as a preload increase. Therefore, the main mechanical stimulus difference lies in the 'phase' – systolic phase for pressure overload and diastolic phase for volume overload, during which the cells sense increased wall stress or stretch. The comparison of physiological changes and cellular responses are reviewed previously [110, 111, 128], and temporal myocardial responses to pressure overload and volume overload are summarized in **Figures 2** and **3**.

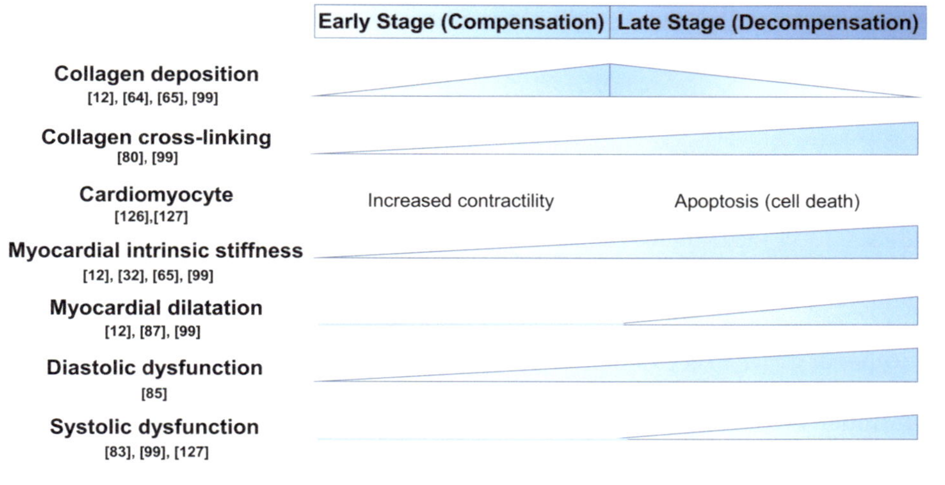

Figure 2.
Temporal changes in myocardial fibrosis, wall stiffening and physiological function in pressure overload induced heart failure progression.

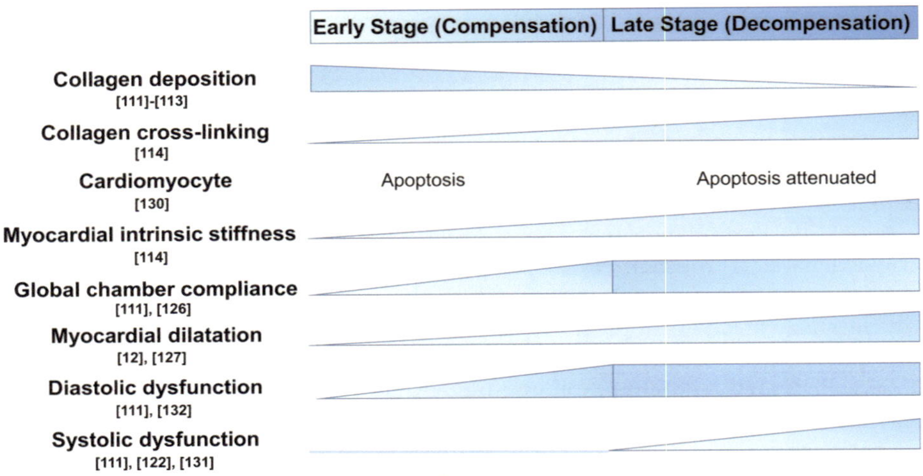

Figure 3.
Temporal changes in myocardial fibrosis, wall stiffening and physiological function in volume overload induced heart failure progression.

The initial (early) remodeling of these two types of overload is different, leading to concentric remodeling (reduced volume and increased wall thickness (h) to radius (r) ratio) in pressure overload and eccentric remodeling (increased volume and reduced h/r ratio) in volume overload. Sometimes these changes are considered as 'adaptive' since the wall stress was normalized by the remodeling (according to Law of Laplace). However, at the cellular level, the 'growth' of cardiomyocytes is in width (pressure overload) and in length (volume overload), respectively. Interestingly, it remains unclear why and how the cardiomyocytes respond to the higher stress at passive and active states differently [129]. Furthermore, as pointed by Pitoulis et al., the late remodeling of these two types of overload 'converges', indicating some common pathways shared in the decompensation stage [128]. The convergence could be caused by a co-existence of these mechanical overloads in the same patient or by the key shared pathways that lead to irreversible remodeling at the late stage of heart failure. These are important questions that remain to be investigated to further improve the clinical management of non-ischemic heart failure patients.

4.4 Biomechanical adaptations in myocardial infarction (ischemia)

Myocardial infarction (MI) is caused by a reduction of blood perfusion in coronary arteries and necrosis—cardiomyocyte cell death [130]. This is the most studied type of heart failure, and there are numerous reviews on ECM changes in MI [1, 2, 21]. Acute and chronic MI lead to significant inflammatory and fibrotic responses by recruitment and activation of neutrophils, macrophages and (myo) fibroblasts. Interstitial and perivascular fibrosis from both replacement and reactive types of collagen accumulation greatly reduces the cardiomyocyte population and muscle contractility. The failure to resolve the fibrotic region (scar) results in a continuous stiffening and dilatation of myocardium and irreversible HF [105]. Compared to the LV, the RV is less susceptible to ischemic injury and it can recover after prolonger coronary occlusion [131, 132]. Therefore, our discussion below is mainly focused on the LV.

The initial response of MI is a cascade of events including the release of MMPs and proteases, which is thought to trigger the post-MI inflammatory response and

DOI: http://dx.doi.org/10.5772/intechopen.96585

to degrade the ECM following proliferative phase [12]. Then, collagen deposition is seen most prevalently in the infarcted region, creating a fibrotic scar [133]. The injured region can be clearly identified, with a high-density mesh of small diameter, aligned collagen fibers [134]. Both collagen types I and III were increased in infarcted LVs [134, 135]. Cleautiens et al. found that type I procollagen expression was increased and peaked by 10-fold four days post-MI, and the elevation remained 90 days after infarction. A similar trend was seen in type III procollagen, however its levels reached a peak later at day 21 post-MI [135]. Alternatively, experimental work also shows that the procollagen type III induction occurs earlier than that of procollagen I [136]. Other fibrillar collagen, such as types V and VI, increased in infarcted areas; for instance, the content of type VI peaked two weeks post-MI [137]. Elastin content was found to increase initially, but it rapidly decreased to values lower than the healthy tissue [134].

Moreover, regional changes occur in the non-infarcted zones. In the rat LV, collagen concentration increased by 50% in the infarcted region, whereas only a 27% increase was reported in the remote region [138]. Furthermore, collagen cross-linking only occurred in the infarcted zone and not in the remote region [138]. Similarly, there was a gradient of changes in the mechanical properties in different regions. The infarcted region showed mechanical alterations to the highest degree; the border zone showed moderate changes, and the remote region showed little to no change [139]. ECM remodeling also arises in the septum and RV with infarcted LV [135]. A 4–5 folds increase in expression of procollagen types I and III mRNA was observed in the septum within three weeks post-MI, followed by a decline to control levels. A less but significant elevation of collagen types I and III also occurred in the RV, with levels of type III remaining elevated 90 days post-MI [135].

The injured area post-MI is predominantly a fibrotic scar with time-dependent changes in mechanical properties. Torres et al. found that the myocardial stiffness increased initially in the MI region, then in the border zone and the remote region. At 28 days post-MI, the myocardial longitudinal stiffness in all three regions decreased, and the reduction was most notable in the MI region with a 40% decrease [139]. Decreases in stiffness in the infarcted zone in late MI have been documented in other studies: the tissue becomes more compliant and more isotropic compared to the healthy tissue [140]. The initial increase in stiffness can be attributed to deposition of collagen and increased collagen isotropy [141]. At later stages, tissue became softer (e.g., 6–8 weeks post-MI) due to collagen degradation [131, 132, 140, 141]. These biomechanical changes are speculated to contribute to the further thinning and dilation of the myocardium leading to irreversible HF. Thus, the end-stage of HF due to MI, pressure overload and volume overload seem to share some common pathways associated with similar biomechanical alterations.

4.5 Biomechanics of heart valves

4.5.1 Biomechanical alterations in heart valve disease

Heart valve disease can occur as a result of various causes such as aging, birth-defects, or infections. Depending on the mechanical abnormalities, the heart valve disease can be categorized into two types—stenosis (with reduced opening of valves) and regurgitation (with incomplete closure of valves). These mechanical changes will increase the mechanical load of the ventricle and reduce cardiac output. Often the disease progression involves a mix of pressure and volume overloads and biventricular dysfunction. Aortic valve disease (AVD) is the most studied heart valve disease and involves high levels of ECM degradation and calcification. For instance, collagen I and III decreases resulted in decreased valve stiffness [142]. But the aggregation of calcium hydroxyapatite (calcification) that replaces degraded collagen can lead to an ultimate stiffening of the valve leaflets. We recommend the reader to these thorough reviews of the biomechanical changes of heart valves in heart valve disease [143–145].

4.5.2 Role of ECM in the mechanics of heart valves

In healthy valve leaflets, collagen makes up ~90% of the ECM and thus is the main load-bearing component. Elastin, proteoglycans, and GAGs also contribute to the mechanical properties of the heart valves. This dense connective tissue is highly organized and present special viscoelastic mechanical behavior (minimal creep but significant stress relaxation) [146]. Therefore, any changes in the collagen fiber orientation or ECM proteins will induce mechanical dysfunction and thus abnormal opening/closing of the valves. The necessity of investigation of valve ECM is recently reviewed [143]. Furthermore, for the atrioventricular valve (i.e., mitral valve or tricuspid valve), the additional connection of the leaflets (LL) to the chordae tendineae (CT) and then further to papillary muscles (PM) has extended the research into transition regions of LL-CT and CT-PM. Because CT rupture is the primary cause of valve regurgitation, the mechanical properties of these regions are critical to delineate the pathology. Advanced methodologies on collagen fiber quantification (e.g., X-ray diffraction) and computational models were recently adopted to investigate the macro- and micro-mechanical properties of these transition regions [147, 148], although the research has been mostly focused on healthy tissues. A nice review of the microstructural mechanical characterization on CT of valves is referred here [149].

Collagen fibers in the CT contribute greatly to the overall function of the valve [147]. From the mitral valve (MV), collagen fibers found in the CT form a 'core' that are oriented longitudinally, with another group of collagen fibers that wrap around the core, offset from the primary axis [150]. Conversely, it has been observed that the CT extending from the tricuspid valve (TV) consists of smaller diameter of fibers but in higher densities than the MV [151]. This difference in collagen formation is a result of the mechanical needs of each CT, as the TV experiences lower loading than the MV. In diseased valves such as myxomatous degeneration of the mitral valve (MDMV), collagen deposition and myofibroblasts activation are observed. Although accumulation of collagen fibers is often associated with increased stiffness, the CT in MDMV actually becomes more compliant. Barber et al. found that the elastic modulus (EM) of healthy CT (132 ± 15 MPa), as well as the failure strength (25.7 ± 1.8 MPa), were significantly higher than those of diseased CT in MDMV (40.4 ± 10.2 MPa and 6.0 ± 0.6 MPa, respectively) [152]. A similar finding was also observed in other valve diseases: myxomatous degeneration of the tricuspid valve by Lim et al. [153], as well as by Casado et al. in the calcified mitral valve CT [154]. In such cases, the collagen fibers in the center core of the CT became disorganized and were no longer formed in tight bundles as they do in healthy CT [153]. Therefore, the altered alignment and dis-organization of the collagen fibers reduce the overall stiffness as well as the tensile strength of the CT.

5. Mechanobiology of cardiac cells

As we discussed above, ECM remodeling is a critical part of tissue's response to altered mechanical loads or other pathological stimulations (e.g., ischemia)

[112–114, 155–160]. This leads to dynamic alterations of the myocardial mechanical properties including the anisotropic, nonlinear elastic behavior as well as the viscoelastic behavior. The clinical relevance of ventricular mechanical behavior has been recently reviewed by our group [76]. In this chapter, we will focus on another impact – the cellular response to substrate biomechanical properties. Because of the prominent role of fibrillar collagen in myocardial mechanical behavior (especially in diseased tissues), we will mainly discuss the cellular response caused by fibrillar collagen deposition. The influence of other ECM proteins on cardiac cells during HF progression is reviewed in these references [12, 161].

The cardiac cell's response to mechanical factors has been mostly investigated by exposing cells to steady or cyclic stretches to mimic myocardial contraction. Compared to the unstretched condition, there was significant increase in procollagen type I activity in cultured fibroblasts under cyclic stretching [156, 157]. Alternatively, Carver et al. evaluated the change in collagen III/I ratio in isolated cardiac fibroblasts. After 12 hours of cyclic loading, there was a 70% increase in the ratio of type III to type I collagen compared to unstretched cells [158]. Not only the collagen synthesis but also the degradation signaling pathway are upregulated by the mechanical loads. Multiple studies have found an upregulation of MMPs, particularly MMP-2 and MMP-9, in response to mechanical stretches compared to unstretched conditions [113, 155, 159]. Since the expression of collagen or procollagen was greater than that of MMP's, a net increase in collagen was observed [155]. This mechanical regulation can be attributed to higher levels of transforming growth factor beta 1 (TGF-β1), which stimulates fibroblast's growth and transdifferentiation and ECM protein synthesis [155, 160]. However, while these findings strongly advocate the mechanical regulation of cardiac fibroblasts and fibrogenesis, the unstretched condition is non-physiological and thus it is difficult to directly translate the findings into pathogenesis of fibrosis and heart failure.

More in-depth investigation of the substrate's mechanical regulation is performed by varying the magnitudes of mechanical strains or comparing the effects of tensile stretch and compression. For example, after 24 hours of culture of cardiac fibroblasts, Lee et al. found that the expression of collagen type I mRNA significantly increased at uniaxial strain of 10% but showed no change at 20% strain. In contrast, type III collagen mRNA expression significantly increased at 10% strain but decreased at 20% strain. The response of collagen type III was more prominent than collagen type I [160]. In compression though, no significant change was noted in type I or III collagen mRNA. The different response between stretch and compression is intriguing as it implies that it is the mechanical signal in diastole (stretching of wall), not in systole (compression of wall), that stimulates the fibrotic response in fibroblasts. But a discrepant result is also reported. Kong et al. observed at compressive cyclic loading (5–20% strain) an upregulation in collagen type I and TGF-β1 expression [155], whereas Lee et al. reported significant increases only in TGF-β1 expression and no notable changes in collagen type I at 6% compressive strain compared to unstretched cells [160].

Besides the mechanical forces, it is also critical to investigate if and how the mechanical stiffness of the substrate affects cellular function. To date, a few *in vitro* studies have reported the regulation of matrix stiffness using synthetic hydrogels tuned to match the stiffness of healthy and diseased myocardial tissues. When cultured on different stiffness of poly (ethylene glycol) (PEG) hydrogels, cardiac fibroblasts were activated into myofibroblasts by increased stiffness (6 vs. 60 kPa) and addition of TGF-β1 stimulation. Furthermore, only the condition medium from fibroblasts cultured on stiff matrix treated with TGF-β1 caused neonatal rat ventricular myocytes enlargement, indicating a synergistic effect of matrix stiffness and TGF-β1 on the activation of myofibroblasts and myocyte hypertrophy [162]. In addition, the fibroblasts responded to the dynamic stiffening of the PEG hydrogel (from 10 to 50 kPa) by increasing cell spread area and reducing nuclei roundness within 2–5 days of culture, mimicking the *in vivo* observation of phenotype changes of fibroblasts [163]. Increased fibroblast cells spreading and collagen type I expression, and decreased collagen III expression were reported in stiffer matrix

Biomechanical Cue	Cell Type	Main Finding	Ref
Polyacrylamide gel with varying stiffnesses: 8 kPa, 15 kPa, 50 kPa, 100 kPa (stiffness range from healthy to diseased heart	Adult Rat Cardiomyocytes (CM)	Contracting force of the CM tends to increase with matrix stiffness. Contractile function, though, is normalized when the matrix stiffness was returned to healthy levels	[165]
PDMS surfaces with varying stiffnesses: 1 kPa, 6 kPa, 20 kPa, 130 kPa (embryonic stiffness to fibrotic heart stiffness)	Adult Rat Cardiomyocytes (CM)	On soft surfaces, CM exhibits greater contraction than those on stiff surfaces. Sarcomeres show faster shortening speeds on soft surfaces. TGF-β1 is more prevalent on stiffer surfaces	[166]
PEG hydrogels with varying stiffnesses: 10 kPa (healthy neonatal heart), 35 kPa (diseased neonatal heart)	Neonatal Rat Ventricular Cardiomyocytes (CM)	Regardless of matrix stiffness, myocytes spontaneously contract and form healthy, organized sarcomeres. However, fractional shortening increased only on soft gels	[167]
Collagen type I matrix morphology (fibrous vs. non-fibrous) and stiffness (48–170 kPa)	Alveolar-like Macrophages	Non-fibrous collagen leads to cells with higher filopodia (actin-rich protrusions) and higher CD206 expression, whereas stiffness has no significant impact on cell behavior	[168]
Fibronectin coated polyacrylamide gels with varying stiffness: soft (1–5 kPa) to represent healthy artery and stiff (280 kPa-70 GPa) to represent atherosclerotic plaque	Human Macrophages	Stiffer substrates leads to larger cell spread area, faster migration speed, less dense but more organized F-actin, and increased proliferation rate	[169]
Polyacrylamide Gel with varying stiffness: 1.2 kPa to represent normal lungs and 150 kPa to represent fibrotic lungs	Mouse macrophages	Cell elasticity increases with increased matrix stiffness; and cell elasticity is a major determinant of innate macrophage function	[170]
PEG-RGD hydrogel with tangent (compressive) modulus from 130 kPa to 840 kPa	Mouse macrophages	Stiffer hydrogel leads to larger cell spread area and more extended branches of actin, activation of pro-inflammatory cytokines, and more severe foreign body response	[171]

Table 3.
Summary of in vitro studies of matrix stiffness dependent changes in cell behavior and function.

(3 vs. 8 kPa), which is postulated to imitate the late stage MI heart with matured scar formation [164]. In the same study, it was also found that the cross-linking of collagen was 'triggered' by non-equibiaxial static stretch mimicking *in vivo* strains in the border region rather than the matrix stiffening. While this experiment design de-couples the mechanical stretch and stiffness, it must be noted that in physiological conditions, these two factors are not independent – a stiffer material will result in reduced stretch/strain under the same pressure. We have listed a few *in vitro* studies that investigated the responses of cardiomyocytes or macrophages to substrate stiffness in **Table 3**. These pilot studies highly support a role of ECM mechanical properties in cellular regulation relevant to cardiac ECM remodeling. Overall, the area of cardiac mechanobiology is still very young, as we reviewed recently [76]. The discrepant elastic moduli reported from literature and different matrix stiffness ranges selected by various groups have increased the difficulty to delineate the cellular responses in the progression of cardiac diseases. Therefore, future studies should aim to further elaborate the role of matrix mechanics (stretch/strain, stiffness, etc.) in heart failure development.

6. Conclusions

The cardiac ECM is critical in maintaining cardiac tissue structure and function. Many studies have been conducted to measure ECM proteins in healthy and diseased myocardium to better understand their roles in cardiac remodeling – including the biomechanical changes. Our review shows that the ECM remodeling (particularly collagen accumulation) in HF is both spatially and temporally dependent. We have compared the myocardial collagen deposition, wall stiffening and systolic and diastolic dysfunction between early and late stages of various types of HF. While the initial remodeling events being quite different among these diseases, common biomechanical changes are shared in the end-stage of HF – ECM degradation with persisted cross-linking, which are associated with thinning and dilatation of the myocardial wall. However, the relation of cardiac fibrosis to the transition from compensation to decompensation remains to be elucidated. Furthermore, we have high-lighted different responses of the LV and RV to 'identical stimulus' (pressure overload, volume overload and ischemia). The interventricular difference should be another important future direction of research, which may help to bring new insights into the pathogenesis and treatment for ventricular failure.

Conflict of interest

The authors declare no conflict of interest.

Author details

Kristen LeBar and Zhijie Wang*
Department of Mechanical Engineering, Colorado State University, Fort Collins, CO, USA

*Address all correspondence to: zhijie.wang@colostate.edu

References

[1] C. Frantz, K. M. Stewart, and V. M. Weaver, "The extracellular matrix at a glance," *Journal of Cell Science*, vol. 123, no. 24. J Cell Sci, pp. 4195-4200, Dec. 15, 2010, doi: 10.1242/jcs.023820.

[2] M. Rienks, A. P. Papageorgiou, N. G. Frangogiannis, and S. Heymans, "Myocardial extracellular matrix: An ever-changing and diverse entity," *Circulation Research*, vol. 114, no. 5. Lippincott Williams & WilkinsHagerstown, MD, pp. 872-888, Feb. 28, 2014, doi: 10.1161/CIRCRESAHA.114.302533.

[3] J. Parvizi and G. . Kim, "Collagen," in *High Yield Orthopaedics*, 2010, pp. 107-109.

[4] J. Rodriguez-Feo, J. Sluijter, D. Kleijn, and G. Pasterkamp, "Modulation of Collagen Turnover in Cardiovascular Disease," Curr. Pharm. Des., vol. 11, no. 19, pp. 2501-2514, Jul. 2005, doi: 10.2174/1381612054367544.

[5] D. A. Duprez *et al.*, "Collagen Turnover Markers in Relation to Future Cardiovascular and Noncardiovascular Disease: The Multi-Ethnic Study of Atherosclerosis," Clin. Chem., vol. 63, no. 7, pp. 1237-1247, Jul. 2017, doi: 10.1373/clinchem.2016.270520.

[6] S. H. Li *et al.*, "Elastin overexpression by cell-based gene therapy preserves matrix and prevents cardiac dilation," J. Cell. Mol. Med., vol. 16, no. 10, pp. 2429-2439, Oct. 2012, doi: 10.1111/j.1582-4934.2012.01560.x.

[7] F. H. Silver, D. Christiansen, P. B. Snowhill, Y. Chen, and W. J. Landis, "The role of mineral in the storage of elastic energy in turkey tendons," Biomacromolecules, vol. 1, no. 2, pp. 180-185, 2000, doi: 10.1021/bm9900139.

[8] B. B. Aaron and J. M. Gosline, "Elastin as a random-network elastomer: A mechanical and optical analysis of single elastin fibers," Biopolymers, vol. 20, no. 6, pp. 1247-1260, Jun. 1981, doi: 10.1002/bip.1981.360200611.

[9] A. J. Cocciolone, J. Z. Hawes, M. C. Staiculescu, E. O. Johnson, M. Murshed, and J. E. Wagenseil, "Elastin, arterial mechanics, and cardiovascular disease," *American Journal of Physiology - Heart and Circulatory Physiology*, vol. 315, no. 2. American Physiological Society, pp. H189–H205, Aug. 01, 2018, doi: 10.1152/ajpheart.00087.2018.

[10] S. Wei, L. T. C. Chow, I. O. L. Shum, L. Qin, and J. E. Sanderson, "Left and right ventricular collagen type I/III ratios and remodeling post-myocardial infarction," J. Card. Fail., vol. 5, no. 2, pp. 117-126, 1999, doi: 10.1016/S1071-9164(99)90034-9.

[11] R. I. Bashey, A. Martinez-Hernandez, and S. A. Jimenez, "Isolation, characterization, and localization of cardiac collagen type VI: Associations with other extracellular matrix components," Circ. Res., vol. 70, no. 5, pp. 1006-1017, 1992, doi: 10.1161/01.res.70.5.1006.

[12] N. G. Frangogiannis, "The extracellular matrix in ischemic and nonischemic heart failure," *Circulation Research*, vol. 125, no. 1. Lippincott Williams and Wilkins, pp. 117-146, Jun. 21, 2019, doi: 10.1161/CIRCRESAHA.119.311148.

[13] X. Shi *et al.*, "Epicardial prestrained confinement and residual stresses: a newly observed heart ventricle confinement interface," J. R. Soc. Interface, vol. 16, no. 152, p. 20190028, Mar. 2019, doi: 10.1098/rsif.2019.0028.

[14] C. Farquharson and S. P. Robins, "The distribution of elastin in developing and adult rat organs using

immunocytochemical techniques," Wiley-Blackwell, 1989. Accessed: Dec. 24, 2020. [Online]. Available: https://www.ncbi.nlm.nih.gov/pmc/articles/PMC1256672/.

[15] V. S. Mujumdar, L. M. Smiley, and S. C. Tyagi, "Activation of matrix metalloproteinase dilates and decreases cardiac tensile strength," Int. J. Cardiol., vol. 79, no. 2-3, pp. 277-286, 2001, doi: 10.1016/S0167-5273(01)00449-1.

[16] N. Latif, P. Sarathchandra, P. M. Taylor, J. Antoniw, and M. H. Yacoub, "Localization and Pattern of Expression of Extracellular Matrix Components in Human Heart Valves," J. Heart Valve Dis., vol. 14, no. 2, pp. 218-227, 2005.

[17] K. Kunzelman, R. Cochran, S. Murphree, W. Ring, E. Verrier, and R. Eberhart, "Differential collagen distribution in the mitral valve and its influence on biomechanical behaviour.," J. Heart Valve Dis., vol. 2, no. 2, pp. 236-244, Mar. 1993.

[18] B. Alberts, A. Johnson, J. Lewis, M. Raff, K. Roberts, and P. Walter, "Fibroblasts and Their Transformations: The Connective-Tissue Cell Family," in *Molecular Biology of the Cell*, 4th ed., 1970.

[19] J. Nüchel *et al.*, "TGFB1 is secreted through an unconventional pathway dependent on the autophagic machinery and cytoskeletal regulators," Autophagy, vol. 14, no. 3, pp. 465-486, Mar. 2018, doi: 10.1080/15548627.2017.1422850.

[20] R. E. Chapman *et al.*, "Matrix metalloproteinase abundance in human myocardial fibroblasts: Effects of sustained pharmacologic matrix metalloproteinase inhibition," J. Mol. Cell. Cardiol., vol. 35, no. 5, pp. 539-548, May 2003, doi: 10.1016/S0022-2828(03)00077-4.

[21] V. Talman and H. Ruskoaho, "Cardiac fibrosis in myocardial infarction—from repair and remodeling to regeneration," *Cell and Tissue Research*, vol. 365, no. 3. Springer Verlag, pp. 563-581, Sep. 01, 2016, doi: 10.1007/s00441-016-2431-9.

[22] N. G. Frangogiannis, "Fibroblasts and the extracellular matrix in right ventricular disease," *Cardiovascular Research*, vol. 113, no. 12. Oxford University Press, pp. 1453-1464, Oct. 01, 2017, doi: 10.1093/cvr/cvx146.

[23] M. L. Coker, M. A. Doscher, C. V. Thomas, Z. S. Galis, and F. G. Spinale, "Matrix metalloproteinase synthesis and expression in isolated LV myocyte preparations," Am. J. Physiol. - Hear. Circ. Physiol., vol. 277, no. 2 46-42, 1999, doi: 10.1152/ajpheart.1999.277.2.h777.

[24] P. Liu, M. Sun, and S. Sader, "Matrix metalloproteinases in cardiovascular disease," *Can. J. Cardiol.*, vol. 22, no. SUPPL. B, pp. 25B-30B, 2006, doi: 10.1016/s0828-282x(06)70983-7.

[25] M. A. Pinkert, R. A. Hortensius, B. M. Ogle, and K. W. Eliceiri, "Imaging the cardiac extracellular matrix," in *Advances in Experimental Medicine and Biology*, vol. 1098, Springer New York LLC, 2018, pp. 21-44.

[26] K. Ozawa, N. Funabashi, H. Takaoka, N. Tanabe, K. Tatsumi, and Y. Kobayashi, "Detection of right ventricular myocardial fibrosis using quantitative CT attenuation of the right ventricular myocardium in the late phase on 320 slice CT in subjects with pulmonary hypertension," Int. J. Cardiol., vol. 228, pp. 165-168, Feb. 2017, doi: 10.1016/j.ijcard.2016.11.131.

[27] E. Spuentrup *et al.*, "Molecular magnetic resonance imaging of myocardial perfusion with EP-3600, a collagen-specific contrast agent initial feasibility study in a swine model," Circulation, vol. 119, no. 13,

pp. 1768-1775, Apr. 2009, doi: 10.1161/CIRCULATIONAHA.108.826388.

[28] S. Shuvaev, E. Akam, and P. Caravan, "Molecular MR Contrast Agents," Invest. Radiol., vol. 56, no. 1, pp. 20-34, 2021.

[29] M. L. Shehata *et al.*, "Myocardial delayed enhancement in pulmonary hypertension: Pulmonary hemodynamics, right ventricular function, and remodeling," *Am. J. Roentgenol.*, vol. 196, no. 1, pp. 87-94, Jan. 2011, doi: 10.2214/AJR.09.4114.

[30] L. Choudhury *et al.*, "Myocardial scarring in asymptomatic or mildly symptomatic patients with hypertrophic cardiomyopathy," J. Am. Coll. Cardiol., vol. 40, no. 12, pp. 2156-2164, Dec. 2002, doi: 10.1016/S0735-1097(02)02602-5.

[31] M. F. Kozak, A. Redington, S. J. Yoo, M. Seed, A. Greiser, and L. Grosse-Wortmann, "Diffuse myocardial fibrosis following tetralogy of Fallot repair: A T1 mapping cardiac magnetic resonance study," Pediatr. Radiol., vol. 44, no. 4, pp. 403-409, 2014, doi: 10.1007/s00247-013-2840-9.

[32] C. Kropidlowski *et al.*, "CMR based measurement of aortic stiffness, epicardial fat, left ventricular myocardial strain and fibrosis in hypertensive patients," *IJC Hear. Vasc.*, vol. 27, Apr. 2020, doi: 10.1016/j.ijcha.2020.100477.

[33] T. Podlesnikar, V. Delgado, and J. J. Bax, "Cardiovascular magnetic resonance imaging to assess myocardial fibrosis in valvular heart disease," Int. J. Cardiovasc. Imaging, vol. 34, no. 1, pp. 97-112, Jan. 2018, doi: 10.1007/s10554-017-1195-y.

[34] A. Protti *et al.*, "Assessment of Myocardial Remodeling Using an Elastin/Tropoelastin Specific Agent with High Field Magnetic Resonance Imaging (MRI)," *J. Am. Heart Assoc.*, vol. 4, no. 8, p. e001851, Aug. 2015, doi: 10.1161/JAHA.115.001851.

[35] D. C. Baumgart *et al.*, "US-based real-time elastography for the detection of fibrotic gut tissue in patients with stricturing Crohn disease," Radiology, vol. 275, no. 3, pp. 889-899, Jun. 2015, doi: 10.1148/radiol.14141929.

[36] X. Qin and B. Fei, "Measuring myofiber orientations from high-frequency ultrasound images using multiscale decompositions," Phys. Med. Biol., vol. 59, no. 14, pp. 3907-3924, Jul. 2014, doi: 10.1088/0031-9155/59/14/3907.

[37] S. Won, C. Davies-Venn, S. Liu, and D. A. Bluemke, "Noninvasive imaging of myocardial extracellular matrix for assessment of fibrosis," *Current Opinion in Cardiology*, vol. 28, no. 3. Curr Opin Cardiol, pp. 282-289, May 2013, doi: 10.1097/HCO.0b013e32835f5a2b.

[38] H. Wahyudi, A. A. Reynolds, Y. Li, S. C. Owen, and S. M. Yu, "Targeting collagen for diagnostic imaging and therapeutic delivery," J. Control. Release, vol. 240, pp. 323-331, Oct. 2016, doi: 10.1016/j.jconrel.2016.01.007.

[39] I. R. Efimov, V. P. Nikolski, and G. Salama, "Optical imaging of the heart," *Circulation Research*, vol. 95, no. 1. Lippincott Williams & Wilkins, pp. 21-33, Jul. 09, 2004, doi: 10.1161/01.RES.0000130529.18016.35.

[40] P. Matteini *et al.*, "Photothermally-induced disordered patterns of corneal collagen revealed by SHG imaging," Opt. Express, vol. 17, no. 6, p. 4868, Mar. 2009, doi: 10.1364/oe.17.004868.

[41] R. Cicchi, N. Vogler, D. Kapsokalyvas, B. Dietzek, J. Popp, and F. S. Pavone, "From molecular structure to tissue architecture:

collagen organization probed by SHG microscopy," J. Biophotonics, vol. 6, no. 2, pp. 129-142, Feb. 2013, doi: 10.1002/jbio.201200092.

[42] V. Crosignani, S. Jahid, A. S. Dvornikov, and E. Gratton, "A deep tissue fluorescence imaging system with enhanced SHG detection capabilities," Microsc. Res. Tech., vol. 77, no. 5, pp. 368-373, 2014, doi: 10.1002/jemt.22354.

[43] D. Hemsley, "Optical Microscopy," Comprehenisve Polym. Sci. Suppl., vol. 1, pp. 765-784, 1989.

[44] W. Schmatz, T. Springer, J. Schelten, and K. Ibel, "Neutron small-angle scattering: experimental techniques and applications," J. Appl. Crystallogr., vol. 7, no. 2, pp. 96-116, Apr. 1974, doi: 10.1107/s0021889874008880.

[45] M. S. Sacks, D. B. Smith, and E. D. Hiester, "A small angle light scattering device for planar connective tissue microstructural analysis," Ann. Biomed. Eng., vol. 25, no. 4, pp. 678-689, 1997, doi: 10.1007/BF02684845.

[46] G. K. Reddy and C. S. Enwemeka, "A simplified method for the analysis of hydroxyproline in biological tissues," Clin. Biochem., vol. 29, no. 3, pp. 225-229, 1996, doi: 10.1016/0009-9120(96)00003-6.

[47] B. Qiu *et al.*, "Measurement of hydroxyproline in collagen with three different methods," *Mol. Med. Rep.*, vol. 10, no. 2, pp. 1157-1163, Aug. 2014, doi: 10.3892/mmr.2014.2267.

[48] H. PUCHTLER and H. ISLER, "THE EFFECT OF PHOSPHOMOLYBDIC ACID ON THE STAINABILITY OF CONNECTIVE TISSUES BY VARIOUS DYES," J. Histochem. Cytochem., vol. 6, no. 4, pp. 265-270, Jul. 1958, doi: 10.1177/6.4.265.

[49] F. Sweat, H. Puchtler, and S. I. Rosenthal, "Sirius Red F3BA as a Stain for Connective Tissue," Arch. Pathol., vol. 68, pp. 69-72, 1964.

[50] Z. Wang, D. A. Schreier, T. A. Hacker, and N. C. Chesler, "Progressive right ventricular functional and structural changes in a mouse model of pulmonary arterial hypertension," *Physiol. Rep.*, vol. 1, no. 7, 2013, doi: 10.1002/phy2.184.

[51] M. Nguyen-Truong *et al.*, "Establishment of adult right ventricle failure in ovine using a graded, animal-specific pulmonary artery constriction model," *Anim. Model. Exp. Med.*, vol. 3, no. 2, pp. 182-192, Jun. 2020, doi: 10.1002/ame2.12124.

[52] M. . Paccinin, "Histology, Verhoeff Stains," *PubMed*, 2020. https://www.ncbi.nlm.nih.gov/books/NBK519050 (accessed Dec. 01, 2020).

[53] F. Tuckett and G. Morriss-Kay, "Alcian Blue staining of glycosaminoglycans in embryonic material: Effect of different fixatives," Histochem. J., vol. 20, no. 3, pp. 174-182, Mar. 1988, doi: 10.1007/BF01746681.

[54] C. Rosenquist *et al.*, "Serum CrossLaps One Step ELISA. First application of monoclonal antibodies for measurement in serum of bone-related degradation products from C-terminal telopeptides of type I collagen," Clin. Chem., vol. 44, no. 11, pp. 2281-2289, 1998.

[55] M. J. Nielsen *et al.*, "The neo-epitope specific PRO-C3 ELISA measures true formation of type III collagen associated with liver and muscle parameter," Am. J. Transl. Res., vol. 5, no. 3, pp. 303-315, 2013.

[56] Z. Wang, D. W. Long, Y. Huang, W. C. W. Chen, K. Kim, and Y. Wang, "Decellularized neonatal cardiac extracellular matrix prevents widespread ventricular remodeling in adult mammals after myocardial

infarction," Acta Biomater., vol. 87, pp. 140-151, Mar. 2019, doi: 10.1016/j.actbio.2019.01.062.

[57] S. S. Virani *et al.*, "Heart disease and stroke statistics—2020 update: A report from the American Heart Association," *Circulation*, vol. 141, no. 9. Lippincott Williams and Wilkins, pp. E139–E596, 2020, doi: 10.1161/CIR.0000000000000757.

[58] C. W. Yancy *et al.*, "2013 ACCF/AHA guideline for the management of heart failure: Executive summary: A report of the American college of cardiology foundation/American Heart Association task force on practice guidelines," *Circulation*, vol. 128, no. 16, pp. 1810-1852, Oct. 2013, doi: 10.1161/CIR.0b013e31829e8807.

[59] A. Dhingra *et al.*, "Epidemiology of Heart Failure with Preserved Ejection Fraction," *Current Heart Failure Reports*, vol. 11, no. 4. Current Science Inc., pp. 354-365, 2014, doi: 10.1007/s11897-014-0223-7.

[60] T. E. Owan, D. O. Hodge, R. M. Herges, S. J. Jacobsen, V. L. Roger, and M. M. Redfield, "Trends in Prevalence and Outcome of Heart Failure with Preserved Ejection Fraction," N. Engl. J. Med., vol. 355, no. 3, pp. 251-259, Jul. 2006, doi: 10.1056/nejmoa052256.

[61] G. Tackling and M. B. Borhade, "Hypertensive Heart Disease," Jun. 2020, Accessed: Dec. 24, 2020. [Online]. Available: https://www.ncbi.nlm.nih.gov/books/NBK539800/.

[62] F. C. Simões *et al.*, "Macrophages directly contribute collagen to scar formation during zebrafish heart regeneration and mouse heart repair," Nat. Commun., vol. 11, no. 1, Dec. 2020, doi: 10.1038/s41467-019-14263-2.

[63] S. Pushpakumar *et al.*, "Angiotensin-II induced hypertension and renovascular remodelling in tissue inhibitor of metalloproteinase 2 knockout mice," J. Hypertens., vol. 31, no. 11, pp. 2270-2281, 2013, doi: 10.1097/HJH.0b013e3283649b33.

[64] J. E. Jalil, C. W. Doering, J. S. Janicki, R. Pick, S. G. Shroff, and K. T. Weber, "Fibrillar collagen and myocardial stiffness in the intact hypertrophied rat left ventricle," Circ. Res., vol. 64, no. 6, pp. 1041-1050, 1989, doi: 10.1161/01.RES.64.6.1041.

[65] D. Chapman, K. T. Weber, and M. Eghbali, "Regulation of fibrillar collagen types I and III and basement membrane type IV collagen gene expression in pressure overloaded rat myocardium.," Circ. Res., vol. 67, no. 4, pp. 787-794, Oct. 1990, doi: 10.1161/01.RES.67.4.787.

[66] M. C. Álvarez *et al.*, "Is cardiac hypertrophy in spontaneously hypertensive rats the cause or the consequence of oxidative stress?," Hypertens. Res., vol. 31, no. 7, pp. 1465-1476, 2008, doi: 10.1291/hypres.31.1465.

[67] K. Yamamoto *et al.*, "Myocardial stiffness is determined by ventricular fibrosis, but not by compensatory or excessive hypertrophy in hypertensive heart," Cardiovasc. Res., vol. 55, no. 1, pp. 76-82, 2002, doi: 10.1016/S0008-6363(02)00341-3.

[68] K. Nagao *et al.*, "Circulating markers of collagen types I, III, and IV in patients with dilated cardiomyopathy: relationships with myocardial collagen expression," *ESC Hear. Fail.*, vol. 5, no. 6, pp. 1044-1051, Dec. 2018, doi: 10.1002/ehf2.12360.

[69] B. C. Berk, K. Fujiwara, and S. Lehoux, "ECM remodeling in hypertensive heart disease," *Journal of Clinical Investigation*, vol. 117, no. 3. J Clin Invest, pp. 568-575, Mar. 01, 2007, doi: 10.1172/JCI31044.

[70] A. J. Woodiwiss *et al.*, "Reduction in myocardial collagen cross-linking

parallels left ventricular dilatation in rat models of systolic chamber dysfunction," Circulation, vol. 103, no. 1, pp. 155-160, Jan. 2001, doi: 10.1161/01.CIR.103.1.155.

[71] M. Kamiya *et al.*, "β3-adrenergic receptor agonist prevents diastolic dysfunction in an angiotensin II-induced cardiomyopathy mouse model," *J. Pharmacol. Exp. Ther.*, p. JPET-AR-2020-000140, Dec. 2020, doi: 10.1124/jpet.120.000140.

[72] J. Fielitz *et al.*, "Activation of the cardiac renin-angiotensin system and increased myocardial collagen expression in human aortic valve disease," J. Am. Coll. Cardiol., vol. 37, no. 5, pp. 1443-1449, Apr. 2001, doi: 10.1016/S0735-1097(01)01170-6.

[73] H. N. Soufen *et al.*, "Collagen content, but not the ratios of collagen type III/I mRNAs, differs among hypertensive, alcoholic, and idiopathic dilated cardiomyopathy," *Brazilian J. Med.* Biol. Res., vol. 41, no. 12, pp. 1098-1104, 2008, doi: 10.1590/S0100-879X2008001200009.

[74] X. Liu, H. Wu, M. Byrne, S. Krane, and R. Jaenisch, "Type III collagen is crucial for collagen I fibrillogenesis and for normal cardiovascular development," Proc. Natl. Acad. Sci. U. S. A., vol. 94, no. 5, pp. 1852-1856, Mar. 1997, doi: 10.1073/pnas.94.5.1852.

[75] M. M. Hoeper and M. Humbert, "The new haemodynamic definition of pulmonary hypertension: Evidence prevails, finally!," *European Respiratory Journal*, vol. 53, no. 3. European Respiratory Society, Mar. 01, 2019, doi: 10.1183/13993003.00038-2019.

[76] W. Liu and Z. Wang, "Current understanding of the biomechanics of ventricular tissues in heart failure," *Bioengineering*, vol. 7, no. 1. MDPI AG, Mar. 01, 2020, doi: 10.3390/bioengineering7010002.

[77] L. Todorovich-Hunter, D. J. Johnson, P. Ranger, and M. Rabinovitch, "Altered Elastin and Collagen Synthesis Associated with Progressive Pulmonary Hypertension Induced by Monocrataline. A Biochemical and Ultrastructural Study," Lab Invest., vol. 58, no. 2, pp. 184-195, 1988.

[78] M. J. Golob, Z. Wang, A. J. Prostrollo, T. A. Hacker, and N. C. Chesler, "Limiting collagen turnover via collagenase-resistance attenuates right ventricular dysfunction and fibrosis in pulmonary arterial hypertension," *Physiol. Rep.*, vol. 4, no. 11, Jun. 2016, doi: 10.14814/phy2.12815.

[79] B. López, A. González, N. Hermida, F. Valencia, E. De Teresa, and J. Díez, "Role of lysyl oxidase in myocardial fibrosis: From basic science to clinical aspects," *American Journal of Physiology - Heart and Circulatory Physiology*, vol. 299, no. 1. Am J Physiol Heart Circ Physiol, Jul. 2010, doi: 10.1152/ajpheart.00335.2010.

[80] A. V. Shinde and N. G. Frangogiannis, "Tissue transglutaminase in the pathogenesis of heart failure," *Cell Death and Differentiation*, vol. 25, no. 3. Nature Publishing Group, pp. 453-456, Mar. 01, 2018, doi: 10.1038/s41418-017-0028-9.

[81] B. López, R. Querejeta, A. González, M. Larman, and J. Díez, "Collagen cross-linking but not collagen amount associates with elevated filling pressures in hypertensive patients with stage C heart failure: Potential role of lysyl oxidase," Hypertension, vol. 60, no. 3, pp. 677-683, Sep. 2012, doi: 10.1161/HYPERTENSIONAHA.112.196113.

[82] B. López *et al.*, "Myocardial collagen cross-linking is associated with heart failure hospitalization in patients with hypertensive heart failure," J. Am. Coll. Cardiol., vol. 67, no. 3, pp. 251-260, Jan. 2016, doi: 10.1016/j.jacc.2015.10.063.

[83] M. R. Hill, M. A. Simon, D. Valdez-Jasso, W. Zhang, H. C. Champion, and M. S. Sacks, "Structural and Mechanical Adaptations of Right Ventricle Free Wall Myocardium to Pressure Overload," Ann. Biomed. Eng., vol. 42, no. 12, pp. 2451-2465, Nov. 2014, doi: 10.1007/s10439-014-1096-3.

[84] S. Donekal *et al.*, "Interstitial fibrosis, left ventricular remodeling, and myocardial mechanical behavior in a population-based multiethnic cohort: The multi-ethnic study of atherosclerosis (mesa) study," Circ. Cardiovasc. Imaging, vol. 7, no. 2, pp. 292-302, 2014, doi: 10.1161/CIRCIMAGING.113.001073.

[85] Å. T. Røe *et al.*, "Increased passive stiffness promotes diastolic dysfunction despite improved Ca2+ handling during left ventricular concentric hypertrophy," Cardiovasc. Res., vol. 113, no. 10, pp. 1161-1172, Aug. 2017, doi: 10.1093/cvr/cvx087.

[86] G. Gilman, T. A. Nelson, W. H. Hansen, B. K. Khandheria, and S. R. Ommen, "Diastolic Function: A Sonographer's Approach to the Essential Echocardiographic Measurements of Left Ventricular Diastolic Function," *Journal of the American Society of Echocardiography*, vol. 20, no. 2. J Am Soc Echocardiogr, pp. 199-209, Feb. 2007, doi: 10.1016/j.echo.2006.08.005.

[87] S. Watanabe *et al.*, "Myocardial stiffness is an important determinant of the plasma brain natriuretic peptide concentration in patients with both diastolic and systolic heart failure," Eur. Heart J., vol. 27, no. 7, pp. 832-838, Apr. 2006, doi: 10.1093/eurheartj/ehi772.

[88] M. R. Zile *et al.*, "Myocardial stiffness in patients with heart failure and a preserved ejection fraction contributions of collagen and titin," Circulation, vol. 131, no. 14, pp. 1247-1259, 2015, doi: 10.1161/CIRCULATIONAHA.114.013215.

[89] D. Vélez-Rendón, E. R. Pursell, J. Shieh, and D. Valdez-Jasso, "Relative Contributions of Matrix and Myocytes to Biaxial Mechanics of the Right Ventricle in Pulmonary Arterial Hypertension," J. Biomech. Eng., vol. 141, no. 9, Sep. 2019, doi: 10.1115/1.4044225.

[90] S. Jang *et al.*, "Biomechanical and hemodynamic measures of right ventricular diastolic function: Translating tissue biomechanics to clinical relevance," *J. Am. Heart Assoc.*, vol. 6, no. 9, Sep. 2017, doi: 10.1161/JAHA.117.006084.

[91] S. Rajdev *et al.*, "Tissue Doppler assessment of longitudinal right and left ventricular strain and strain rate in pulmonary artery hypertension," Echocardiography, vol. 23, no. 10, pp. 872-879, Nov. 2006, doi: 10.1111/j.1540-8175.2006.00337.x.

[92] M. E. M. De Siqueira *et al.*, "Characterization and clinical significance of right ventricular mechanics in pulmonary hypertension evaluated with cardiovascular magnetic resonance feature tracking," J. Cardiovasc. Magn. Reson., vol. 18, no. 1, p. 39, Jun. 2016, doi: 10.1186/s12968-016-0258-x.

[93] P. Trip *et al.*, "Clinical relevance of right ventricular diastolic stiffness in pulmonary hypertension," Eur. Respir. J., vol. 45, no. 6, pp. 1603-1612, Jun. 2015, doi: 10.1183/09031936.00156714.

[94] M. Murayama *et al.*, "Simple and noninvasive method to estimate right ventricular operating stiffness based on echocardiographic pulmonary regurgitant velocity and tricuspid annular plane movement measurements during atrial contraction," *Int. J. Cardiovasc. Imaging*, vol. 35, no. 10, pp. 1871-1880, Oct. 2019, doi: 10.1007/s10554-019-01637-2.

[95] S. Rain *et al.*, "Right ventricular diastolic impairment in patients with pulmonary arterial hypertension," *Circulation*, vol. 128, no. 18, pp. 2016-2025, Oct. 2013, doi: 10.1161/CIRCULATIONAHA.113.001873.

[96] R. Avazmohammadi, M. Hill, M. Simon, and M. Sacks, "Transmural remodeling of right ventricular myocardium in response to pulmonary arterial hypertension," *APL Bioeng.*, vol. 1, no. 1, Dec. 2017, doi: 10.1063/1.5011639.

[97] M. K. Friedberg and A. N. Redington, "Right versus left ventricular failure: Differences, similarities, and interactions," Circulation, vol. 129, no. 9, pp. 1033-1044, Mar. 2014, doi: 10.1161/CIRCULATIONAHA.113.001375.

[98] W. Liu *et al.*, "Correlations between the right ventricular passive elasticity and organ function in adult ovine," *J. Integr. Cardiol.*, vol. 6, pp. 1-6, 2020, doi: 10.15761/JIC.1000294.

[99] M. Jankowich, S. A. Abbasi, A. Vang, and G. Choudhary, "Right ventricular fibrosis is related to pulmonary artery stiffness in pulmonary hypertension: A cardiac magnetic resonance imaging study," *American Journal of Respiratory and Critical Care Medicine*, vol. 200, no. 6. American Thoracic Society, pp. 776-779, Sep. 15, 2019, doi: 10.1164/rccm.201903-0580LE.

[100] J. Díez, R. Querejeta, B. López, A. González, M. Larman, and J. L. Martínez Ubago, "Losartan-dependent regression of myocardial fibrosis is associated with reduction of left ventricular chamber stiffness in hypertensive patients," Circulation, vol. 105, no. 21, pp. 2512-2517, May 2002, doi: 10.1161/01.CIR.0000017264.66561.3D.

[101] K. Masada, S. Miyagawa, Y. Sakai, A. Harada, T. Kanaya, and Y. Sawa, "Synthetic Prostacyclin Agonist Attenuates Pressure-Overloaded Cardiac Fibrosis by Inhibiting FMT," Mol. Ther. - Methods Clin. Dev., vol. 19, pp. 210-219, Dec. 2020, doi: 10.1016/j.omtm.2020.09.005.

[102] N. Varo *et al.*, "Losartan inhibits the post-transcriptional synthesis of collagen type I and reverses left ventricular fibrosis in spontaneously hypertensive rats," J. Hypertens., vol. 17, no. 1, pp. 107-114, 1999, doi: 10.1097/00004872-199917010-00016.

[103] C. G. Brilla, R. C. Funck, and H. Rupp, "Lisinopril-mediated regression of myocardial fibrosis in patients with hypertensive heart disease," Circulation, vol. 102, no. 12, pp. 1388-1393, Sep. 2000, doi: 10.1161/01.CIR.102.12.1388.

[104] K. T. Weber, Y. Sun, I. C. Gerling, and R. V. Guntaka, "Regression of Established Cardiac Fibrosis in Hypertensive Heart Disease," *American Journal of Hypertension*, vol. 30, no. 11. Oxford University Press, pp. 1049-1052, Nov. 01, 2017, doi: 10.1093/ajh/hpx054.

[105] H. J. Bogaard and N. F. Voelkel, "Is myocardial fibrosis impairing right heart function?," *American Journal of Respiratory and Critical Care Medicine*, vol. 199, no. 12. American Thoracic Society, pp. 1458-1459, Jun. 15, 2019, doi: 10.1164/rccm.201812-2307ED.

[106] S. Crnkovic *et al.*, "Disconnect between fibrotic response and right ventricular dysfunction," Am. J. Respir. Crit. Care Med., vol. 199, no. 12, pp. 1550-1560, Jun. 2019, doi: 10.1164/rccm.201809-1737OC.

[107] M. K. Friedberg and A. N. Redington, "Right versus left ventricular failure: Differences, similarities, and interactions," Circulation, vol. 129, no. 9, pp. 1033-1044, Mar. 2014, doi: 10.1161/CIRCULATIONAHA.113.001375.

[108] R. J. Wiesner, H. Ehmke, J. Faulhaber, R. Zak, and J. C. Ru¨egg, "Dissociation of Left Ventricular Hypertrophy, β-Myosin Heavy Chain Gene Expression, and Myosin Isoform Switch in Rats After Ascending Aortic Stenosis," *Circulation*, vol. 95, no. 5, pp. 1253-1259, Mar. 1997, doi: 10.1161/01.CIR.95.5.1253.

[109] C. E. Simpson and P. M. Hassoun, "Myocardial fibrosis as a potential maladaptive feature of right ventricle remodeling in pulmonary hypertension," *American Journal of Respiratory and Critical Care Medicine*, vol. 200, no. 6. American Thoracic Society, pp. 662-663, Sep. 15, 2019, doi: 10.1164/rccm.201906-1154ED.

[110] A. Diwan and G. W. Dorn, "Decompensation of cardiac hypertrophy: Cellular mechanisms and novel therapeutic targets," *Physiology*, vol. 22, no. 1. Physiology (Bethesda), pp. 56-64, Feb. 2007, doi: 10.1152/physiol.00033.2006.

[111] G. W. Dorn, "The fuzzy logic of physiological cardiac hypertrophy," *Hypertension*, vol. 49, no. 5. Hypertension, pp. 962-970, May 2007, doi: 10.1161/HYPERTENSIONAHA.106.079426.

[112] T. D. Ryan *et al.*, "Left Ventricular Eccentric Remodeling and Matrix Loss Are Mediated by Bradykinin and Precede Cardiomyocyte Elongation in Rats With Volume Overload," J. Am. Coll. Cardiol., vol. 49, no. 7, pp. 811-821, Feb. 2007, doi: 10.1016/j.jacc.2006.06.083.

[113] D. B. Murray, J. D. Gardner, G. L. Brower, and J. S. Janicki, "Effects of nonselective endothelin-1 receptor antagonism on cardiac mast cell-mediated ventricular remodeling in rats," *Am. J. Physiol. - Hear. Circ. Physiol.*, vol. 294, no. 3, Mar. 2008, doi: 10.1152/ajpheart.00622.2007.

[114] M. Ruzicka, F. W. Keeley, and F. H. H. Leenen, "The renin-angiotensin system and volume overload-induced changes in cardiac collagen and elastin," *Circulation*, vol. 90, no. 4 I, pp. 1989-1996, 1994, doi: 10.1161/01.CIR.90.4.1989.

[115] D. S. Iimoto, J. W. Covell, and E. Harper, "Increase in cross-linking of type I and type III collagens associated with volume-overload hypertrophy," Circ. Res., vol. 63, no. 2, pp. 399-408, 1988, doi: 10.1161/01.RES.63.2.399.

[116] R. C. Childers, I. Sunyecz, T. A. West, M. J. Cismowski, P. A. Lucchesi, and K. J. Gooch, "Role of the cytoskeleton in the development of a hypofibrotic cardiac fibroblast phenotype in volume overload heart failure," Am. J. Physiol. - Hear. Circ. Physiol., vol. 316, no. 3, pp. H596–H608, Mar. 2019, doi: 10.1152/ajpheart.00095.2018.

[117] J. S. Janicki *et al.*, "Cardiac mast cell regulation of matrix metalloproteinase-related ventricular remodeling in chronic pressure or volume overload," *Cardiovascular Research*, vol. 69, no. 3. Cardiovasc Res, pp. 657-665, Feb. 15, 2006, doi: 10.1016/j.cardiores.2005.10.020.

[118] Y. wen Chen *et al.*, "Dynamic molecular and histopathological changes in the extracellular matrix and inflammation in the transition to heart failure in isolated volume overload," *Am. J. Physiol. - Hear. Circ. Physiol.*, vol. 300, no. 6, Jun. 2011, doi: 10.1152/ajpheart.01104.2010.

[119] L. Cohen *et al.*, "Cardiac remodeling secondary to chronic volume overload is attenuated by a novel MMP9/2 blocking antibody," PLoS One, vol. 15, no. 4, Apr. 2020, doi: 10.1371/journal.pone.0231202.

[120] Y. J. H. J. Taverne, A. Sadeghi, B. Bartelds, A. J. J. C. Bogers, and D.

Merkus, "Right ventricular phenotype, function, and failure: a journey from evolution to clinics," *Heart Failure Reviews*. Springer, 2020, doi: 10.1007/s10741-020-09982-4.

[121] S. Reddy *et al.*, "Physiologic and molecular characterization of a murine model of right ventricular volume overload," *Am. J. Physiol. Circ. Physiol.*, vol. 304, no. 10, pp. H1314–H1327, May 2013, doi: 10.1152/ajpheart.00776.2012.

[122] T. Pham, C. M. Zgierski-Johnston, K. Tran, A. J. Taberner, D. S. Loiselle, and J. C. Han, "Energy expenditure for isometric contractions of right and left ventricular trabeculae over a wide range of frequencies at body temperature," Sci. Rep., vol. 9, no. 1, Dec. 2019, doi: 10.1038/s41598-019-45273-1.

[123] E. C. El Hajj, M. C. El Hajj, V. K. Ninh, and J. D. Gardner, "Inhibitor of lysyl oxidase improves cardiac function and the collagen/MMP profile in response to volume overload," *Am. J. Physiol. - Hear. Circ. Physiol.*, vol. 315, no. 3, pp. H463–H473, Sep. 2018, doi: 10.1152/ajpheart.00086.2018.

[124] K. L. Herrmann, A. D. McCulloch, and J. H. Omens, "Glycated collagen cross-linking alters cardiac mechanics in volume-overload hypertrophy," Am. J. Physiol. - Hear. Circ. Physiol., vol. 284, no. 4 53-54, Apr. 2003, doi: 10.1152/ajpheart.00168.2002.

[125] J. L. Emery and J. H. Omens, "Mechanical regulation of myocardial growth during volume-overload hypertrophy in the rat," Am. J. Physiol. - Hear. Circ. Physiol., vol. 273, no. 3 42-43, 1997, doi: 10.1152/ajpheart.1997.273.3.h1198.

[126] J. Harper, E. Harper, and J. W. Covell, "Collagen characterization in volume-overload- and pressure-overload- induced cardiac hypertrophy in minipigs," Am. J. Physiol. - Hear. Circ. Physiol., vol. 265, no. 2 34-32, 1993, doi: 10.1152/ajpheart.1993.265.2.h434.

[127] G. L. Brower and J. S. Janicki, "Contribution of ventricular remodeling to pathogenesis of heart failure in rats," 2001. [Online]. Available: http://www.ajpheart.org.

[128] F. G. Pitoulis and C. M. Terracciano, "Heart Plasticity in Response to Pressure- and Volume-Overload: A Review of Findings in Compensated and Decompensated Phenotypes," *Frontiers in Physiology*, vol. 11. Frontiers Media S.A., Feb. 13, 2020, doi: 10.3389/fphys.2020.00092.

[129] Fiorillo, C., Nediani, C., Ponziani, V., Giannini, L., Celli, A., Nassi, N., Formigli, L., Perna, A. M., & Nassi, P. (2005). Cardiac volume overload rapidly induces oxidative stress-mediated myocyte apoptosis and hypertrophy. *Biochimica et Biophysica Acta - Molecular Basis of Disease*, 1741(1-2), 173-182. https://doi.org/10.1016/j.bbadis.2005.03.015

[130] M. F. Berry *et al.*, "Mesenchymal stem cell injection after myocardial infarction improves myocardial compliance," *Am. J. Physiol. - Hear. Circ. Physiol.*, vol. 290, no. 6, Jun. 2006, doi: 10.1152/ajpheart.01017.2005.

[131] J. A. Goldstein, "Acute Right Ventricular Infarction: Insights for the Interventional Era," Curr. Probl. Cardiol., vol. 37, no. 12, pp. 533-557, Dec. 2012, doi: 10.1016/j.cpcardiol.2012.05.001.

[132] J. A. Goldstein, "Pathophysiology and management of right heart ischemia," *Journal of the American College of Cardiology*, vol. 40, no. 5. Elsevier Inc., pp. 841-853, Sep. 04, 2002, doi: 10.1016/S0735-1097(02)02048-X.

[133] M. Dobaczewski, M. Bujak, P. Zymek, G. Ren, M. L. Entman, and N. G. Frangogiannis, "Extracellular

matrix remodeling in canine and mouse myocardial infarcts," Cell Tissue Res., vol. 324, no. 3, pp. 475-488, Jun. 2006, doi: 10.1007/s00441-005-0144-6.

[134] K. P. Quinn *et al.*, "Optical metrics of the extracellular matrix predict compositional and mechanical changes after myocardial infarction," Sci. Rep., vol. 6, Nov. 2016, doi: 10.1038/srep35823.

[135] J. P. Cleautiens, M. J. Verluyten, and M. J. Daemen, "Collagen remodeling after myocardial infarction in the rat heart," Am. J. Pathol., vol. 147, no. 2, pp. 325-338, 1995.

[136] K. Inoue, S. Kusachi, K. Niiya, Y. Kajikawa, and T. Tsuji, "Sequential changes in the distribution of type I and III collagens in the infarct zone: Immunohistochemical study of experimental myocardial infarction in the rat," Coron. Artery Dis., vol. 6, no. 2, pp. 153-158, 1995, doi: 10.1097/00019501-199502000-00010.

[137] T. Ishizu *et al.*, "Left ventricular strain and transmural distribution of structural remodeling in hypertensive heart disease," Hypertension, vol. 63, no. 3, pp. 500-506, Mar. 2014, doi: 10.1161/HYPERTENSIONAHA.113.02149.

[138] R. J. McCormick, T. I. Musch, B. C. Bergman, and D. P. Thomas, "Regional differences in LV collagen accumulation and mature cross-linking after myocardial infarction in rats," Am. J. Physiol. - Hear. Circ. Physiol., vol. 266, no. 1 35-31, 1994, doi: 10.1152/ajpheart.1994.266.1.h354.

[139] W. M. Torres *et al.*, "Regional and temporal changes in left ventricular strain and stiffness in a porcine model of myocardial infarction," *Am. J. Physiol. - Hear. Circ. Physiol.*, vol. 315, no. 4, pp. H958–H967, Oct. 2018, doi: 10.1152/ajpheart.00279.2018.

[140] S. Wang *et al.*, "Biomechanical assessment of myocardial infarction using optical coherence elastography," Biomed. Opt. Express, vol. 9, no. 2, p. 728, Feb. 2018, doi: 10.1364/boe.9.000728.

[141] T. E. Raya, R. G. Gay, L. Lancaster, M. Aguirre, C. Moffett, and S. Goldman, "Serial changes in left ventricular relaxation and chamber stiffness after large myocardial infarction in rats," Circulation, vol. 77, no. 6, pp. 1424-1431, 1988, doi: 10.1161/01.CIR.77.6.1424.

[142] V. K. Krishnamurthy, A. M. Opoka, C. B. Kern, F. Guilak, D. A. Narmoneva, and R. B. Hinton, "Maladaptive matrix remodeling and regional biomechanical dysfunction in a mouse model of aortic valve disease," Matrix Biol., vol. 31, no. 3, pp. 197-205, Apr. 2012, doi: 10.1016/j.matbio.2012.01.001.

[143] K. M. Kodigepalli *et al.*, "Biology and biomechanics of the heart valve extracellular matrix," *Journal of Cardiovascular Development and Disease*, vol. 7, no. 4. MDPI AG, pp. 1-22, Dec. 01, 2020, doi: 10.3390/jcdd7040057.

[144] D. Wiltz *et al.*, "Extracellular Matrix Organization, Structure, and Function," in *Calcific Aortic Valve Disease*, InTech, 2013.

[145] M. S. Sacks and J. Liao, Eds., *Advances in Heart Valve Biomechanics: Valvular Physiology, Mechanobiology, and Bioengineering*. Cham: Springer International Publishing, 2018.

[146] Z. Wang, M. J. Golob, and N. C. Chesler, "Viscoelastic Properties of Cardiovascular Tissues," in *Viscoelastic and Viscoplastic Materials*, InTech, 2016.

[147] R. S. Madhurapantula, G. Krell, B. Morfin, R. Roy, K. Lister, and J. P. R. O. Orgel, "Advanced Methodology and Preliminary Measurements of Molecular and Mechanical Properties of Heart Valves under Dynamic Strain," Int. J. Mol. Sci., vol. 21, no. 3, p. 763, Jan. 2020, doi: 10.3390/ijms21030763.

[148] R. Roy *et al.*, "Functional Grading of a Transversely Isotropic Hyperelastic Model with Applications in Modeling Tricuspid and Mitral Valve Transition Regions," Int. J. Mol. Sci., vol. 21, no. 18, p. 6503, Sep. 2020, doi: 10.3390/ijms21186503.

[149] C. J. Ross, J. Zheng, L. Ma, Y. Wu, and C.-H. Lee, "Mechanics and Microstructure of the Atrioventricular Heart Valve Chordae Tendineae: A Review," Bioengineering, vol. 7, no. 1, p. 25, Mar. 2020, doi: 10.3390/bioengineering7010025.

[150] K. O. Lim and D. R. Boughner, "Scanning electron microscopical study of human mitral valve chordae tendineae," Arch. Pathol. Lab Med, vol. 101, no. 5, pp. 236-238, 1977.

[151] K. O. LIM, "Mechanical Properties and Ultrastructure of Normal Human Tricuspid Valve Chordae Tendineae," Jpn. J. Physiol., vol. 30, no. 3, pp. 455-464, 1980, doi: 10.2170/jjphysiol.30.455.

[152] J. . Barber, N. . Ratliff, D. . Cosgrove, B. . Griffin, and I. Vesely, "Myxomatous mitral valve chordae. I: Mechanical properties," J Hear. Valve Dis, vol. 10, no. 3, pp. 320-324, 2001.

[153] K. O. Lim, D. R. Boughner, and D. G. Perkins, "Ultrastructure and Mechanical Properties of Chordae Tendineae from a Myxomatous Tricuspid Valve," Jpn. Heart J., vol. 24, no. 4, pp. 539-548, 1983, doi: 10.1536/ihj.24.539.

[154] J. A. Casado *et al.*, "Determination of the mechanical properties of normal and calcified human mitral chordae tendineae," J. Mech. Behav. Biomed. Mater., vol. 13, pp. 1-13, Sep. 2012, doi: 10.1016/j.jmbbm.2012.03.016.

[155] M. Kong *et al.*, "Cardiac Fibrotic Remodeling on a Chip with Dynamic Mechanical Stimulation," *Adv. Healthc. Mater.*, vol. 8, no. 3, Feb. 2019, doi: 10.1002/adhm.201801146.

[156] G. E. Lindahl *et al.*, "Activation of fibroblast procollagen α1(I) transcription by mechanical strain is transforming growth factor-β-dependent and involves increased binding of CCAAT-binding factor (CBF/NF-Y) at the proximal promoter," J. Biol. Chem., vol. 277, no. 8, pp. 6153-6161, Feb. 2002, doi: 10.1074/jbc.M108966200.

[157] J. Atance, M. J. Yost, and W. Carver, "Influence of the extracellular matrix on the regulation of cardiac fibroblast behavior by mechanical stretch," J. Cell. Physiol., vol. 200, no. 3, pp. 377-386, Sep. 2004, doi: 10.1002/jcp.20034.

[158] W. Carver, M. L. Nagpal, M. Nachtigal, T. K. Borg, and L. Terracio, "Collagen expression in mechanically stimulated cardiac fibroblasts.," Circ. Res., vol. 69, no. 1, pp. 116-122, Jul. 1991, doi: 10.1161/01.RES.69.1.116.

[159] R. A. García, K. L. Brown, R. S. Pavelec, K. V. Go, J. W. Covell, and F. J. Villarreal, "Abnormal cardiac wall motion and early matrix metalloproteinase activity," Am. J. Physiol. - Hear. Circ. Physiol., vol. 288, no. 3 57-53, Mar. 2005, doi: 10.1152/ajpheart.00860.2004.

[160] A. A. Lee, T. Delhaas, A. D. McCulloch, and F. J. Villarreal, "Differential responses of adult cardiac fibroblasts to in vitro biaxial strain patterns," J. Mol. Cell. Cardiol., vol. 31, no. 10, pp. 1833-1843, 1999, doi: 10.1006/jmcc.1999.1017.

[161] N. G. Frangogiannis and J. C. Kovacic, "Extracellular Matrix in Ischemic Heart Disease, Part 4/4: JACC Focus Seminar," *Journal of the American College of Cardiology*, vol. 75, no. 17. Elsevier USA, pp. 2219-2235, May 05, 2020, doi: 10.1016/j.jacc.2020.03.020.

[162] T. L. Ceccato *et al.*, "Defining the Cardiac Fibroblast Secretome in a Fibrotic Microenvironment," *J.*

Am. Heart Assoc., vol. 9, no. 19, p. e017025, Oct. 2020, doi: 10.1161/JAHA.120.017025.

[163] K. A. Günay *et al.*, "PEG–Anthracene Hydrogels as an On-Demand Stiffening Matrix To Study Mechanobiology," *Angew. Chemie - Int. Ed.*, vol. 58, no. 29, pp. 9912-9916, Jul. 2019, doi: 10.1002/anie.201901989.

[164] K. M. Herum, J. Choppe, A. Kumar, A. J. Engler, and A. D. McCulloch, "Mechanical regulation of cardiac fibroblast profibrotic phenotypes," Mol. Biol. Cell, vol. 28, no. 14, pp. 1871-1882, Jul. 2017, doi: 10.1091/mbc.E17-01-0014.

[165] E. D. van Deel *et al.*, "In vitro model to study the effects of matrix stiffening on Ca2+ handling and myofilament function in isolated adult rat cardiomyocytes," J. Physiol., vol. 595, no. 14, pp. 4597-4610, Jul. 2017, doi: 10.1113/JP274460.

[166] P. Pandey *et al.*, "Cardiomyocytes Sense Matrix Rigidity through a Combination of Muscle and Non-muscle Myosin Contractions," *Dev. Cell*, vol. 44, no. 3, pp. 326-336.e3, Feb. 2018, doi: 10.1016/j.devcel.2017.12.024.

[167] W. Wan, K. K. Bjorkman, E. S. Choi, A. L. Panepento, K. S. Anseth, and L. A. Leinwand, "Cardiac myocytes respond differentially and synergistically to matrix stiffness and topography," *bioRxiv*. bioRxiv, p. 682930, Jun. 27, 2019, doi: 10.1101/682930.

[168] G. F. Vasse *et al.*, "Collagen morphology influences macrophage shape and marker expression in vitro," J. Immunol. Regen. Med., vol. 1, pp. 13-20, Mar. 2018, doi: 10.1016/j.regen.2018.01.002.

[169] K. M. Adlerz, H. Aranda-Espinoza, and H. N. Hayenga, "Substrate elasticity regulates the behavior of human monocyte-derived macrophages," Eur. Biophys. J., vol. 45, no. 4, pp. 301-309, May 2016, doi: 10.1007/s00249-015-1096-8.

[170] R. B. Patel *et al.*, "Diffuse right ventricular fibrosis in heart failure with preserved ejection fraction and pulmonary hypertension," *ESC Hear. Fail.*, vol. 7, no. 1, pp. 253-263, Feb. 2020, doi: 10.1002/ehf2.12565.

[171] A. K. Blakney, M. D. Swartzlander, and S. J. Bryant, "The effects of substrate stiffness on the in vitro activation of macrophages and in vivo host response to poly(ethylene glycol)-based hydrogels," *J. Biomed. Mater. Res. - Part A*, vol. 100 A, no. 6, pp. 1375-1386, Jun. 2012, doi: 10.1002/jbm.a.34104.

Chapter 3

The Extracellular Matrix of the Human and Whale Cornea and Sclera: Implications in Glaucoma and Other Pathologies

Elena Vecino, Noelia Ruzafa, Xandra Pereiro, Ane Zulueta, Alfredo Sarmiento and Alejandro Díez

Abstract

The cornea is the transparent part of the eye that allows light to enter into the eye and reach the retina, thereby activating the neurons that will send messages to the brain. The sclera is the hard-white part of the eye, and its main function is to provide structure and form to the eye, and to support the retina. Indeed, while the cornea best performs its main functions when transparent and it is capable of adapting its curvature to allow the eye to focus, the sclera must be opaque and hard to function correctly. Both structures are mainly composed of collagen, some elastic fibres and ground substance, all components of the Extracellular Matrix. The disposition of the collagen fibres and the amount of ground substance around the fibres is responsible for the differences in the aspect of both these structures. In this chapter, for the first time we have compared the structure and ultrastructure of the cornea and sclera in humans and the whale adult (18mts) *Balaenoptera physalus*, the second largest animal on the planet. We will discuss how the differences in their structure may be related to the maintenance of intraocular pressure in their distinct environments, which is of particular clinical interest as increased intraocular pressure is one of the main causes underlying the development of open angle glaucoma.

Keywords: cornea, sclera, extracellular matrix, structure, ultrastructure, collagen, whale, eyes

1. Introduction

The cornea and sclera are the two most external structures of the eye and the extracellular matrix (ECM) plays a crucial role in their activity. While the cornea is transparent and located in the front the eye, the sclera is the white part that forms the rest of the eye, giving it its spherical form, and providing hardness and structural protection to the internal part of the eye. Different types of collagen constitute the core of both structures, which is surrounded by the so-called "ground substance" that lies between the collagen fibers and around the few cells that are present in these elements. The viscoelastic properties of the cornea and sclera define the distensibility of the eye, which is related to the control these structures exert over intraocular pressure. Moreover, the cornea fulfils its main functions at the interface of the eye with air or water (depending on the habitat). Indeed, the cornea is the principal refracting surface of the dioptric system of the eye, which is why it is transparent, avascular, viscoelastic and quite resistant to deformation.

The structural and chemical composition of the ECM of very large eyes like the whale's eye has been little studied. The human eye measures approximately 2.3 cm in diameter and the two whale's eyes that we have analyzed were 12 and 13 cm in diameter (**Figure 1**). In this chapter we compare the morphological and structural aspects of the human cornea and sclera with those structures in one of the largest eyes ever studied. We consider that at least some of the differences in the structure of these eyes is likely to help adapt the whale to its very extreme conditions of life. These animals live between two very different habitats, capable of rapidly shifting between the water surface and the very deep sea, experiencing huge changes in pressure that their eye can only support without deforming thanks to the strong structure of both the cornea and sclera.

Here we will consider the two structures separately, the cornea and sclera, although there is a continuation between both in the eye. The composition of both structures is very similar, mainly comprised of an ECM that contains collagen, as mentioned above, although the organization of the collagen fibers in each differs underlies their distinct viscoelastic characteristics. Quick-freezing and the deep-etching methods have been used in ultrastructural studies of the collagen fibers in the cornea and sclera, demonstrating that corneal collagen fibers were separated by moderate interfibrillar spaces. By contrast, scleral collagen fibers were organized compactly, with fewer interconnecting filaments. In the sclera, the collagen fibers have a wider diameter (around 200 nm) than those in the cornea (around 40 nm), and the periodicity of the collagen striations was variable in each structure, although in the sclera these striations were difficult to detect because of the surrounding ground substance [1]. Here we used several techniques to study the whale's cornea and sclera, from classical histochemical trichromic staining (**Figure 2**), fluorescent light microscopy (**Figure 3**) to scanning electron microscopy (SEM) (**Figures 4** and **5**), in addition to Raman spectroscopy (**Figure 6**). While microscopy will enable us to determine the structure and ultrastructure of the tissue, Raman spectroscopy is a technique that can be used to optically probe the molecular changes in the tissue. The result of this technique is a spectrum

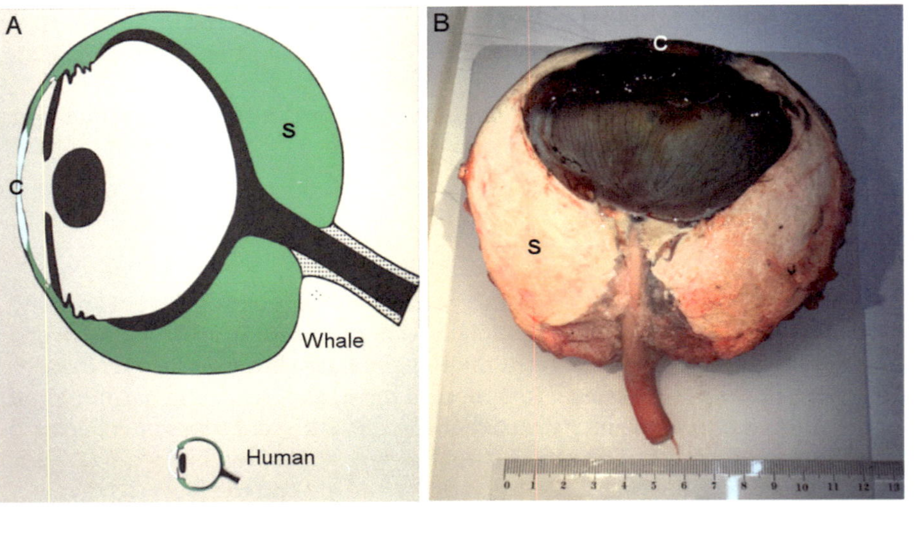

Figure 1.
(A) Scheme of the whale and human eye in proportional scale. In green the sclera and in blue the cornea (B) Picture of half of the whale's eye. (C) cornea, (S) Sclera.

characterized by shifts in wave numbers, which in many cases can be associated with the vibration of particular chemical bond (or single functional group) in the molecule [2]. We will describe the two structures cornea and sclera, comparing human and whale main differences.

2. Methods

Four methods were used to study the extracellular matrix. Two light microscopy methods, for that purpose the cornea and sclera were fixed for 12 h with paraformaldehyde (PAF) 4% and for the other two techniques of electron microscopy a post fixation with 2,5% glutaraldehyde for 2 h was performed after the previous fixation with PAF. The first histological technique used was Masson's trichrome staining, performed in 5 micrometers paraffin sections to visualize the collagen fibers in blue/green from the extracellular matrix (**Figure 2**). The second technique used was fluorescence microscopic technique, to determine the organization of the keratocytes. For that purpose the nuclei of the keratocytes were stained with DAPI in cryostat sections (14 micrometers) (**Figure 3**). The third technique was the scanning electron microscopy (SEM) to visualize the ultrastructure of the matrix components, for that purpose, small portions (few mm^2) of cornea and sclera were dehydrated in increasing gradation of alcohol followed by complete dehydration with hexamethyldisilazane (HDMS), then the pieces were oriented in the platform

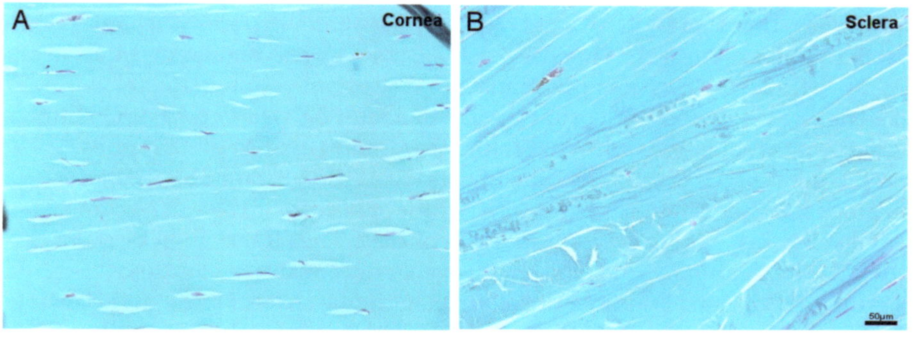

Figure 2.
Trichrome staining of paraffin sections from cornea (A) and sclera (B) of whale's eye. Note the linear and parallel organization of the fibres in cornea and the different orientation of the fibres in the sclera. Scale bar 50 µm.

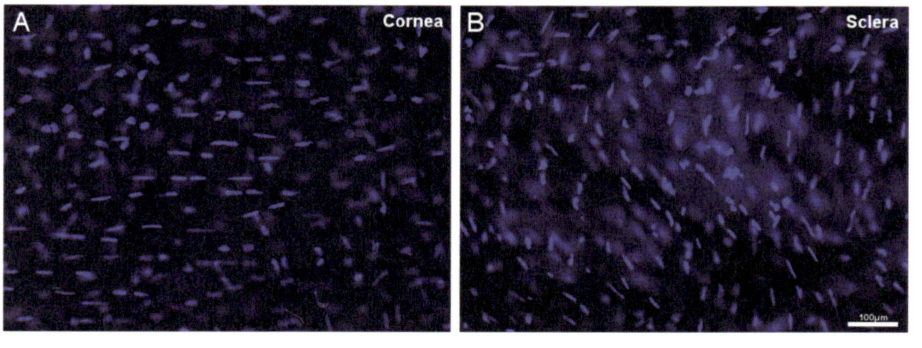

Figure 3.
Fluorescence microscopic picture of the nuclei of the keratocytes. Note their distribution and orientation in cornea (A) and sclera (B) stained with DAPI. In the cornea there are lower number of keratocytes and they are more organized than in the sclera. Scale bar 100 µm.

of the microscope and coated with gold (**Figures 4** and **5**). The last technique was RAMAN microscopy, for that purpose small portions of cornea and sclera were dehydrated as for the SEM, but the HDMS step was not carried out. Thus, the samples were analyzed with a confocal InVia Raman (Renishaw) connected to a spectrophotometer and an excitation laser of 785 nm was connected to a Leica microscope to register the spectro of the different tissues (**Figure 6**).

3. The cornea

As indicated above, the cornea is a transparent organ that allows the light to enter into the eye. The features that contribute to its transparency are a thin epithelium, the absence of blood vessels and its chemical composition, mainly comprised of collagen and some important ground substance, with very few cells. The cornea has five main parts: (1) the epithelium; (2) Bowman's layer; (3) the stroma; (4) Descement's layer; and (5) the endothelium. In this chapter we will concentrate on the stroma, of which the ECM is the main component.

In humans the cornea is approximately 0.5 mm thick, while in the whale's eye it measures 3 to 4 millimetres. In both cases it is composed almost entirely of collagenous lamellae. The collagen fibres are organized in lamellae approximately 6 mm in diameter but with certain variability in their width and thickness. The lamellae are arranged parallel to the corneal surface and sometimes they form loose fibrillar networks. The collagen fibres within the bundles lie parallel to each other, and they are uniform in size and spacing, a feature produced by the cementing ground substance that is distributed regularly between the fibres (**Figure 2**). In the most peripheral cornea, the lamellae gradually adopt a less regular orientation and little-by-little their structure approximates to the organization in the sclera [3]. The collagen fibres in the central cornea vary in diameter between 21–65 nm in humans [4], data that is consistent with that found in our human SEM preparations.

A few specialized fibroblasts called keratocytes can be found between the collagen fibres, and they are responsible for the synthesis of the collagen and ground substance. Only a small proportion of the cornea is occupied by these cells, around 2–3% in humans, and as such, it is generally considered an almost acellular structure [5]. The small number of cells present in the corneal stroma, the avascular nature of this structure and the very well-organized collagen lamellae, all contributes to the characteristic transparency of the cornea (**Figures 2** and **3**).

The ground substance in the cornea consists of mucoproteins, glycoproteins and other substances exclusive to collagen, and it forms a cement like filling in the space between the corneal fibres. In 1969, using alkaline lead, citrate and uranyl acetate staining, 2 nm diameter filaments were seen to exist at right angles to the collagen fibres that they connected, postulating that these were the proteoglycans that bind to the corneal collagen D-period [6]. Using cationic dyes (alcian blue, cuprolinic blue, cupromeronic blue) in a critical electrolyte mode, the presence of proteoglycans was confirmed. Later studies described these proteoglycans to be keratan sulphate (lumincan) and dermatan sulphate (decorin) in the cornea [7], while only dermatan sulphate proteoglycan was found in the sclera, bound to the same sites as in cornea [8]. It was subsequently proposed that these molecules play a role in maintaining the relative positions of the fibrils, which is important for corneal transparency [9, 10]. So far, the most sophisticated and less invasive technique to study the ultrastructure of the cornea, without affecting the physiological state of hydration is the X-ray [11] and this will help in the future for a better understand the pathophysiology of the cornea.

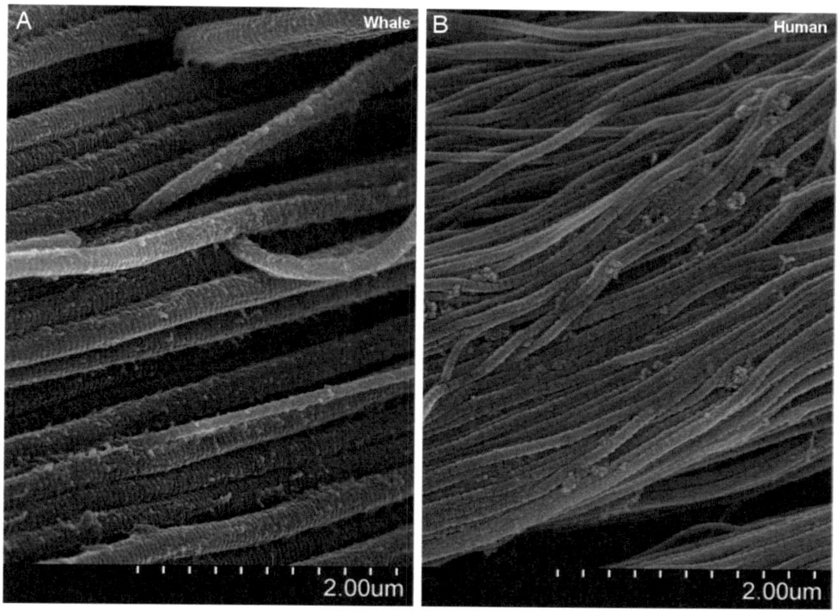

Figure 4.
Scanning electron microscopy pictures of the whale's (A) and human (B) corneas. Note the distribution of the fibers in laminar bundles. Scale in μm.

The corneas of the whale studied are oval in shape, with axes of 5 x 3 cm, and they have a convex outer surface. The corneal thickness varies between the centre, where it is 2.5 mm thick, and the periphery where it is thicker, measuring 4 mm at the corneal-scleral boundary. The diameter of the corneal collagen fibres also differed significantly between the human and whale. Thus, while in humans the corneal fibres are around 60 nm in diameter, in whale they measure around 200 nm (**Figure 4**). The composition of the collagen is probably very similar in both species, not least because their histochemical staining is very similar, also resembling that of the pig, rat and mouse cornea. Moreover, and in addition to SEM and TEM, when the whale cornea and sclera was studied by Raman spectroscopy, the characteristics of the peaks for the collagen components were similar to those in humans [12].

4. The sclera

The sclera is the white part of the eye and it is relative thin, ranging from 0.6 mm in the anterior part to 1 mm in the posterior part of human eyes. However, the sclera is very thick in large whales like the fin whales that we have studied, and it measures 3 to 4 cm at the back of the eye, although it is thinner (0,5 cm) in the anterior part (**Figure 1**). This thick and quite hard structure serves as a coffer in which the sensitive parts of the eye like the retina can be protected from the intense pressures these animals are exposed to when swimming in the deep seas.

The ECM of the human sclera is mainly composed of type I, III, V and VI collagen. The principal function of type I collagen is to resist tension, while type III collagen is considered essential in maintain the structure of expandable organs and type V collagen has been implicated in controlling fibril diameter. Type V collagen also fulfils a role in anchoring to the basement membrane and adjacent stromal matrix, a function it shares with type VI collagen [13]. In the sclera, the collagen fibrils have various diameters, ranging from 25 to 230 nm. Although these collagen fibrils form bundles, their arrangement is more heterogeneous in the sclera than in the cornea. These collagen bundles vary in width and thickness, often sprouting branches and intertwining with each other, at least in humans [14]. Moreover, in the sclera there is a narrower interfibrillar distance than in the cornea and the ground substance is more abundant, impairing the discrimination of the band periodicity of the collagen fibres. Indeed, it has been necessary to use special treatments and atomic force microscopy to describe the differences in the periodicity of the collagen bands between the cornea and sclera [15].

In transverse section of the eye the human sclera is thinner towards the corneo-scleral junction, while it thickens in the medial direction, posterior to the vitreous chamber, where it joins the bundle of the optic nerve. The dorsal part of the sclera is larger than the ventral domain, which means that the optic nerve can exit the eye with a ventral disposition. The collagen fibres that make up the sclera are mainly embedded in the ground substance and the characterization of the different types of collagen fibres has been achieved in humans by immunogold EM staining [16]. The fibres are tightly packed and arranged in different directions, which provides the eyeball with strength and shape (**Figures 2B** and **5**). Close to the corneo-scleral limbus, large blood vessels circulate not far from the angle, forming a ring. In the sclero-corneal stroma of the limbus there is a large number of pigmented cells and numerous channels are present in this area that form the well-developed trabecular meshwork responsible for draining aqueous humour toward the veins.

The analysis of the whale's sclera using Raman spectrometry showed us that even when the thick sclera is quite hard (with a texture like a spongiosum bone), hydroxyapatite does not appear to be present and thus, we concluded that the hard sclera is not ossified Indeed, when comparing the spectrometry fingerprint of human bone with that from the whale's sclera, both structures share collagen peaks (**Figure 6**). The sclera is likely to be important in preserving the shape of the eyeball, shielding it from the effects of the deforming forces. Indeed, this large eye can be retracted or protruded thanks to a large muscle that surrounds the optic nerve

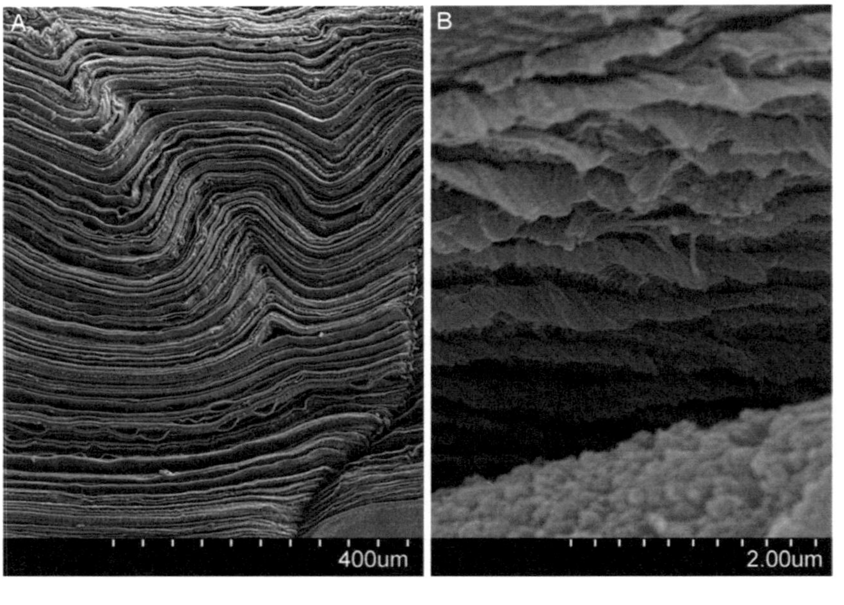

Figure 5.
Scanning electron microscopic pictures of (A) and (B) are both sclera collagen fibers from whale's sclera. (A) lower and (B) higher magnification. Scale in μm.

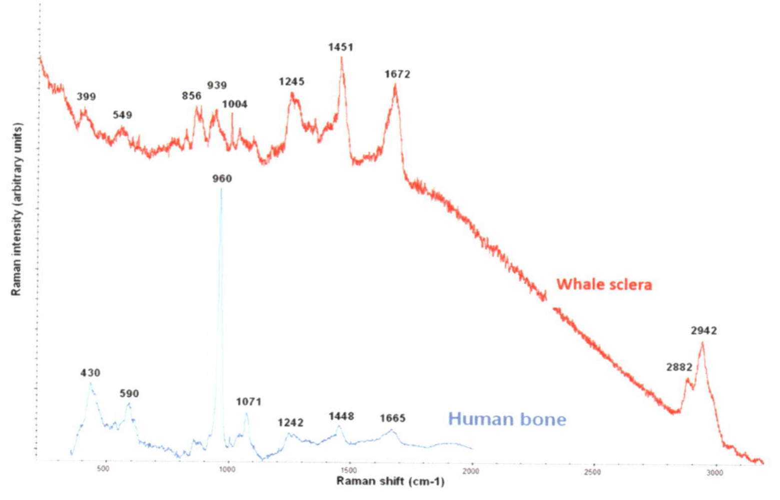

Figure 6.
Raman spectrometry (A) overlapping the spectra from human bone (blue) and whale's sclera.

and that is full of blood vessels, the ophthalmic rete [17]. It is possible that this large muscle also helps the eye and a thick sclera resist the pressure of the deep seas and avoid eye deformation.

5. Implications of the cornea and sclera extracellular matrix in glaucoma and other eye pathologies

Glaucoma is the main cause of blindness in the world. Although there are several types of glaucoma, the most common is characterized by an increase in the intraocular pressure (IOP) that induces neurodegeneration in the retina. Indeed, glaucoma leads to the death of the retinal ganglion cells that are the responsible for sending visual information from the eye to the brain, thus causing blindness [18]. The increasing of the intraocular pressure is due to the elevated secretion of aqueous humour or to a reduction in the evacuation of it, mainly through the trabecular meshwork. So far, in the human eye it has not been detected any sensor to detect and control the intraocular pressure. Interestingly, encapsulated sensory corpuscles are specialized nerve endings located in the corneo-scleral area that do not have a very clear function. These have been found in different cetaceans and in the whale *Balaenoptera acutorostrata*, they were also found in the buccal cavity. It has been speculated that these corpuscles might play a role in detecting and controlling the pressure in different areas of the eye including the sclera [19].

The thickness of the cornea is very important and has to be taken into consideration in order to measure IOP correctly. Since the way to measure the IOP is through the cornea, the instruments used must be adapted to the mean cornea thickness. However, in order to correct the defects, a refraction technique has been developed that involves correcting the curvature of the cornea by reshaping the stroma of the cornea with a laser, LASIK surgery. The thickness of the cornea is critical to be able to perform this surgery, particularly since the mean cornea thickness in humans is 500 um and it reaches a maximum of 600 um, and LASIK surgery should not be performed on thinner corneas. After LASIK surgery, the patient should retain a minimum of 250 μm corneal thickness. In this sense, IOP measurements can vary depending on the thickness of the cornea, being underestimated in patients with thinner corneas and overestimated in patients with thicker corneas. Another side effect of re-shaping a thin cornea is the deformation in the central part, which can alter corneal curvature, so-called keratoconus. This is a phenomenon that leads to a gradually bulging of the cornea outwards into a cone shape, which causes blurry, distorted vision. In order to correct this keratoconus crosslinking of the collagen fibers should be performed by applying UV light to the collagen fibers, thereby reinforcing the structure of the cornea. The UV light together with the application of riboflavin (vitamin B2) will enhance the bonds between collagen fibers in the stroma of the cornea [20]. It is also hypothesized that stiffening of ocular structures, including cornea and sclera may be related to the pathogenesis of glaucoma [21].

Another alteration to the cornea that can influence IOP measurement is the prolonged use of contact lenses. Initially, contact lenses can induce a flattening of the cornea during the first months of use, but prolonged use can cause a thinning of the cornea with some deformation. Thus, we can conclude that prolonged use of contact lenses negatively influences corneal physiology. Aging can also change central and peripheral corneal thickness. By using ultrasonic pachymetry in 250 patients aged 9 to 97 years, it was concluded that central corneal thickness increases significantly with age, whereas the degree of symmetry decreases [22]. Accordingly, there are different factors that can affect corneal thickness and thus, IOP measurements, which could influence the detection and treatment of glaucoma.

The sclera provides a tough fibrous support structure for the retina and optic nerve, fulfilling a biomechanical function that may be crucial in glaucoma. Several studies have assessed collagen fiber architecture in order to identify if uniaxial (one preferred direction) or biaxial (two directions) collagen organization of the sclera is related to glaucoma. So far, changes in fiber orientation have been detected between glaucomatous and non-glaucomatous eyes, although it could be an adaptation to the elevated pressure and it is not clear if there is a predisposition to glaucomatous axon damage [23]. However, the very hard, strong and thick sclera present in the whale's eye means there is no capacity for distension or structural modification. As such, any elevation in IOP in whales would be sensed by the retina. The other structure in the eye that is sensitive to IOP is the lamina cribosa (LC) or cribiform plate that forms a scaffold for the passage of the optic nerve's axon bundles, anchoring the bundles to each other and to the sides forming the optic nerve. It reinforces the posterior eye, protecting it from injury at the site of optic nerve exit. The LC is subject to mechanical strain as it lies at the border between two different compartments subject to pressure: the anterior compartment to IOP and the retrobulbar compartment to that of the cerebrospinal fluid [24]. Hence, the LC has been proposed as the main site controlling the pressure that represents the insult to retinal ganglion cell axons in glaucoma [25]. Moreover, the LC thickness and the posterior displacement of its components have been associated with the rate of progressive retinal fiber layer thinning and the severity of glaucoma. Changes in the structure of the LC have been found in patients with glaucoma, indicating that these structural changes could provide information regarding the evolution of glaucoma [26]. However, in our large exanimated animals, the LC of the whales is as hard as the sclera, which means it will be very difficult for it to deform. Thus, in these animals there is a very limited possibility for the eye to deform in response changes in the IOP.

6. Conclusions

In conclusion, we have evaluated the structure of the eye in the second largest mammalian on the planet, the long fin whale, considering the possible functional consequences of its features. These eyes are around 150 times larger than the human eye, although their structure is very similar and their ECM components are also comparable, albeit in different proportions. Thus, the cornea and sclera are thicker, adapting to the whale's ecosystem and to the physiology of their body size. The very large structures and the rigid ECM lead us to consider the implication of the ECM in eye diseases like glaucoma and keratoconus, which in these animals will be very difficult to explain in the context of their very distinct dimensions and structure.

Acknowledgements

We would like to thank the support by: Grupos UPV/EHU GIU18/50; PIBA 2020-1-0026; Retos MINECO FEDER (PID2019-111139RB-I00); FECYT-19-14532 to EV. Basque Government Post-Doctoral Fellowship to XP.

Conflict of interest

The authors declare no conflict of interest.

Notes/thanks/other declarations

We would like to thank AMBAR association for the help in getting the whale's eye. In addition, we would like to thank all persons that unconditionally encouraged and supported us to carry out the research and to show the results in the art exhibition of scientific photography about the eye of the whale in the Bizkaia Aretoa in February 2020.

Author details

Elena Vecino[1*], Noelia Ruzafa[1], Xandra Pereiro[1], Ane Zulueta[1], Alfredo Sarmiento[2] and Alejandro Díez[3]

1 Cell Biology and Histology Department, Faculty of Medicine and Nursery, Experimental-Ophthalmo-Biology Group (www.ehu.es/gobe), University of Basque Country UPV/EHU, Leioa, Spain

2 Coupled Multispectroscopy Singular Laboratory (Raman-Lapsea) (SGIker), Spain

3 General Service for Analytical and High Resolution Microscopy in Biomedicine (SGIker), Spain

*Address all correspondence to: elena.vecino@ehu.es

References

[1] Yamabayashi S, Ohno S, Aguilar RN, Furuya T, Hosoda M, Tsukahara S. Ultrastructural studies of collagen fibers of the cornea and sclera by a quick-freezing and deep-etching method. Ophthalmic Res. 1991;23(6):320-329. DOI: 10.1159/000267129

[2] Petrich W. Mid-Infrared and Raman Spectroscopy for Medical Diagnostics. Applied Spectroscopy Reviews. 2001;36:181-237.

[3] Hogan MJ, Alvarado JA, Weddell JE. Histology of the human eye. Saunders, editor 1971.

[4] Kayes J, Holmberg A. The fine structure of Bowman's layer and the basement membrane of the corneal epithelium. Am J Ophthalmol. 1960;50:1013-1021. DOI: 10.1016/0002-9394(60)90354-8

[5] Maurice DM, Riley MV. Biochemistry of the cornea. In: Graymore CN, editor. The Biochemisty of the eye New York: Academic Press Inc.; 1968.

[6] Smith JW, Frame J. Observations on the collagen and proteinpolysaccharide complex of rabbit cornea stroma. J Cell Sci. 1969;4(2):421-436.

[7] Scott JE, Haigh M. Identification of specific binding sites for keratan sulphate proteoglycans and chondroitin-dermatan sulphate proteoglycans on collagen fibrils in cornea by the use of cupromeronic blue in 'critical-electrolyte-concentration' techniques. Biochem J. 1988;253(2):607-610. DOI: 10.1042/bj2530607

[8] Young RD (1985). The ultrastructural organization of proteoglycans and collagen in human and rabbit scleral matrix. J. Cell Sci. 74:95-104

[9] Borcherding MS, Blacik LJ, Sittig RA, Bizzell JW, Breen M, Weinstein HG. Proteoglycans and collagen fibre organization in human corneoscleral tissue. Exp Eye Res. 1975;21(1):59-70. DOI: 10.1016/0014-4835(75)90057-3

[10] Scott JE, Thomlinson AM. The structure of interfibrillar proteoglycan bridges (shape modules') in extracellular matrix of fibrous connective tissues and their stability in various chemical environments. J Anat. 1998;192 (Pt 3):391-405. DOI: 10.1046/j.1469-7580.1998.19230391.x

[11] Meek KM, Boote C. The use of X-ray scattering techniques to quantify the orientation and distribution of collagen in the corneal stroma. Prog Retin Eye Res. 2009;28(5):369-392. DOI: 10.1016/j.preteyeres.2009.06.005

[12] Garcia Martínez M, Bullock AJ, MacNeil S, Rehman IU. Characterisation of structural changes in collagen with Raman Spectroscopy. Applied Spectroscopy Reviews. 2019;54(6):509-542.

[13] Birk DE, Fitch JM, Linsenmayer TF. Organization of collagen types I and V in the embryonic chicken cornea. Invest Ophthalmol Vis Sci. 1986;27(10):1470-1477.

[14] Komai Y, Ushiki T. The three-dimensional organization of collagen fibrils in the human cornea and sclera. Invest Ophthalmol Vis Sci. 1991;32(8):2244-2258.

[15] Miyagawa A, Kobayashi M, Fujita Y, Hamdy O, Hirano K, Nakamura M, et al. Surface ultrastructure of collagen fibrils and their association with proteoglycans in human cornea and sclera by atomic force microscopy and energy-filtering transmission electron microscopy. Cornea. 2001;20(6):651-656. DOI: 10.1097/00003226-200108000-00019

[16] Marshall GE, Konstas AG, Lee WR. Collagens in the aged human macular sclera. Curr Eye Res. 1993;12(2):143-153. DOI: 10.3109/02713689308999482

[17] Dawson WW. The cetacean eye. In: Herman LM, editor. Cetacean behavior: Mechanisms and functions. New york: Wiley; 1980. p. 53-100.

[18] Jonas JB, Aung T, Bourne RR, Bron AM, Ritch R, Panda-Jonas S. Glaucoma. Lancet. 2017;390(10108):2183-2193. DOI: 10.1016/S0140-6736(17)31469-1

[19] Miller S, Samuelson D, Dubielzig R. Anatomic features of the cetacean globe. Vet Ophthalmol. 2013;16 Suppl 1:52-63. DOI: 10.1111/vop.12025

[20] Wollensak G, Spoerl E, Seiler T. Riboflavin/ultraviolet-a-induced collagen crosslinking for the treatment of keratoconus. Am J Ophthalmol. 2003;135(5):620-627. DOI: 10.1016/s0002-9394(02)02220-1

[21] Liu B, McNally S, Kilpatrick JI, Jarvis SP, O'Brien CJ. Aging and ocular tissue stiffness in glaucoma. Surv Ophthalmol. 2018;63(1):56-74. DOI: 10.1016/j.survophthal.2017.06.007

[22] Constantini E, Touzeau O, Gaujoutx T, Baslim E, Kopito R, Borderie VM, et al. Age-Related Changes in Central and Peripheral Corneal Thickness. ARVO Annual Meeting Abstract2009.

[23] Pijanka JK, Coudrillier B, Ziegler K, Sorensen T, Meek KM, Nguyen TD, et al. Quantitative mapping of collagen fiber orientation in non-glaucoma and glaucoma posterior human sclerae. Invest Ophthalmol Vis Sci. 2012;53(9):5258-5270. DOI: 10.1167/iovs.12-9705

[24] Wilczek M. The Lamina Cribrosa and Its Nature. Br J Ophthalmol. 1947;31(9):551-565. DOI: 10.1136/bjo.31.9.551

[25] Yan DB, Coloma FM, Metheetrairut A, Trope GE, Heathcote JG, Ethier CR. Deformation of the lamina cribrosa by elevated intraocular pressure. Br J Ophthalmol. 1994;78(8):643-648. DOI: 10.1136/bjo.78.8.643

[26] Tan NYQ, Tham YC, Thakku SG, Wang X, Baskaran M, Tan MCL, et al. Changes in the Anterior Lamina Cribrosa Morphology with Glaucoma Severity. Sci Rep. 2019;9(1):6612. DOI: 10.1038/s41598-019-42649-1

Chapter 4

The Evolutionary Origin of Elastin: Is Fibrillin the Lost Ancestor?

Fernando Rodriguez-Pascual

Abstract

Elastin is the extracellular matrix protein providing large arteries, lung parenchyma and skin with the properties of extensibility and elastic recoil. Within these tissues, elastin is found as a polymer formed by tropoelastin monomers assembled and cross-linked. In addition to specific protein regions supporting the covalent cross-links, tropoelastin is featured by the presence of highly repetitive sequences rich in proline and glycine making up the so-called hydrophobic domains. These protein segments promote structural flexibility and disordered protein properties, a fundamental aspect to explain its elastomeric behavior. Unlike other matrix proteins such as collagens or laminins, elastin emerged relatively late in evolution, appearing at the divergence of jawed and jawless fishes, therefore present in all species from sharks to humans, but absent in lampreys and other lower chordates and invertebrates. In spite of an intense interrogation of the key aspects in the evolution of elastin, its origin remains still elusive and an ancestral protein that could give rise to a primordial elastin is not known. In this chapter, I review the main molecular features of tropoelastin and the available knowledge on its evolutionary history as well as establish hypotheses for its origin. Considering the remarkable similarities between the hydrophobic domains of the first recognizable elastin gene from the elasmobranch *Callorhinchus milii* with certain fibrillin regions from related fish species, I raise the possibility that fibrillins might have provided protein domains to an ancestral elastin that thereafter underwent significant evolutionary changes to give the elastin forms found today.

Keywords: elastin, evolution, fibrillin, cross-linking, extracellular matrix

1. Introduction

Elastin is an extracellular matrix (ECM) component of tissues such as the large arteries and lung parenchyma, among others. Even if it is considerably less abundant compared to other matrix proteins, such as collagens, it impacts the biomechanical properties as it is ultimately responsible for the extensibility and elastic recoil [1]. Different aspects make elastin an unusual protein, for example, its molecular structure, fundamental to understand its function, or the complexity of the mechanisms giving to the formation of elastin-based polymers in the ECM. In the genomics era, where uncountable genomes are decoded and extraordinary valuable information on the phylogenetic relationships has been established, it is remarkable that the evolutionary origin of elastin remains unclear. This fact looks anomalous considering that its roots are relatively close to us in an evolutionary scale, compared to other cell components originated at the dawn of life. It is worthwhile and the main objective of this work to review the main molecular features of elastin and the current knowledge on its evolutionary history as well as to explore new avenues to explain its origin. A couple of hypotheses are put on the table to stimulate (and provoke) discussion and further research. In this regard, I can only recognize the contributions of Professor Fred W. Keely to the overall understanding of the biology of elastin, and particularly to its evolutionary relationships [2].

2. Molecular features of tropoelastin

Most ECM components such as collagens or fibronectin are usually large polypeptides with numerous domains that include repetitive motifs allowing multimerization into supramolecular assemblies [3]. Tropoelastin, the monomer making up polymeric elastin, features some of these characteristics but is indeed an unusual ECM protein. While fibril-forming collagens such as types I and II contain almost 1500 residues and fibronectin goes far beyond 2000, tropoelastin rarely exceeds 800 aminoacids, with some species such as the dog harboring elastin chains of about 500 residues. Moreover, crystal structures of particular domains of collagens or fibronectin have been determined by X-ray diffraction whereas those of elastin have persistently remained elusive [4, 5]. This is in fact a consequence of the disordered nature of the elastin polypeptide, a key feature to understand its elastomeric properties [6, 7]. Analysis of all known elastins shows an alternation between lysine-rich and hydrophobic domains. The former contain the lysine residues destined to take part in the covalent cross-linking catalyzed by members of the lysyl oxidase (LOX) family, therefore named cross-linking domains. The latter are consistently rich in hydrophobic domains and have been shown to be essential for the elastomeric behavior [1]. With noticeable variations, these domains contain stretches of the sequence VPGVG in multiple combinations either forming repeats or included within glycine/proline-rich segments. Molecular dynamic simulations and experimental studies have shown that the hydrophobic motifs support the disordered state by promoting structural flexibility [6, 8, 9]. In fact, this plasticity is likely not limited to the hydrophobic domains but also impacts the whole molecule, considering the fact that lysine oxidation and further condensation of cross-linking domains occur in a random manner, as recently reported [10]. The *quasi*-stochastic assembly of highly flexible tropoelastin monomers results in aggregates that exhibit the elastomeric properties and are the basis for the formation of elastic fibers in numerous tissues, including the lung, skin and blood vessels.

3. Evolutionary history of elastin

It may be easily inferred that this sophisticated natural material has required thousands of millions of years to be shaped by the evolution. However, unlike collagens or laminins, whose origin dates back 770–880 million years (Myr) to the emergence of the Metazoa, tropoelastin appeared late on stage [2, 11]. In fact, not even it emerged with the blood vascular systems, a feature of vertebrates and many invertebrates, but it made its debut 400 Myr ago with the jawed vertebrates, therefore absent in jawless fishes such as the lamprey and hagfish. Elastin has been repeatedly invoked as the vascular component that allowed the development of closed circulatory systems. However, these are present in a wide variety of invertebrates including annelids, cephalopods and non-vertebrate chordates [12]. Here, the magic word is blood pressure. It is actually the presence of elastin that led to closed systems exhibiting high pressures (from 30 to 200 mmHg) in jawed vertebrates, in contrast to non-elastin based systems in invertebrates and lower vertebrates, with blood pressures values ranging from one or a few mmHg up to 20–30 mmHg. Most ancient elastin so far reported includes that from the elasmobranch fish *Callorhinchus milii* (elephant shark) and displays the characteristic alternating pattern of hydrophobic and cross-linking domains (**Figure 1**). Amphibians such as the western clawed frog (*Xenopus tropicalis*) or teleost fishes such as zebrafish (*Danio rerio*) and fugu (*Takifugu rubripes*) feature two versions of tropoelastins, *elastin a* (*elna*) and *elastin b* (*elnb*), compared to other vertebrate genomes that possesses

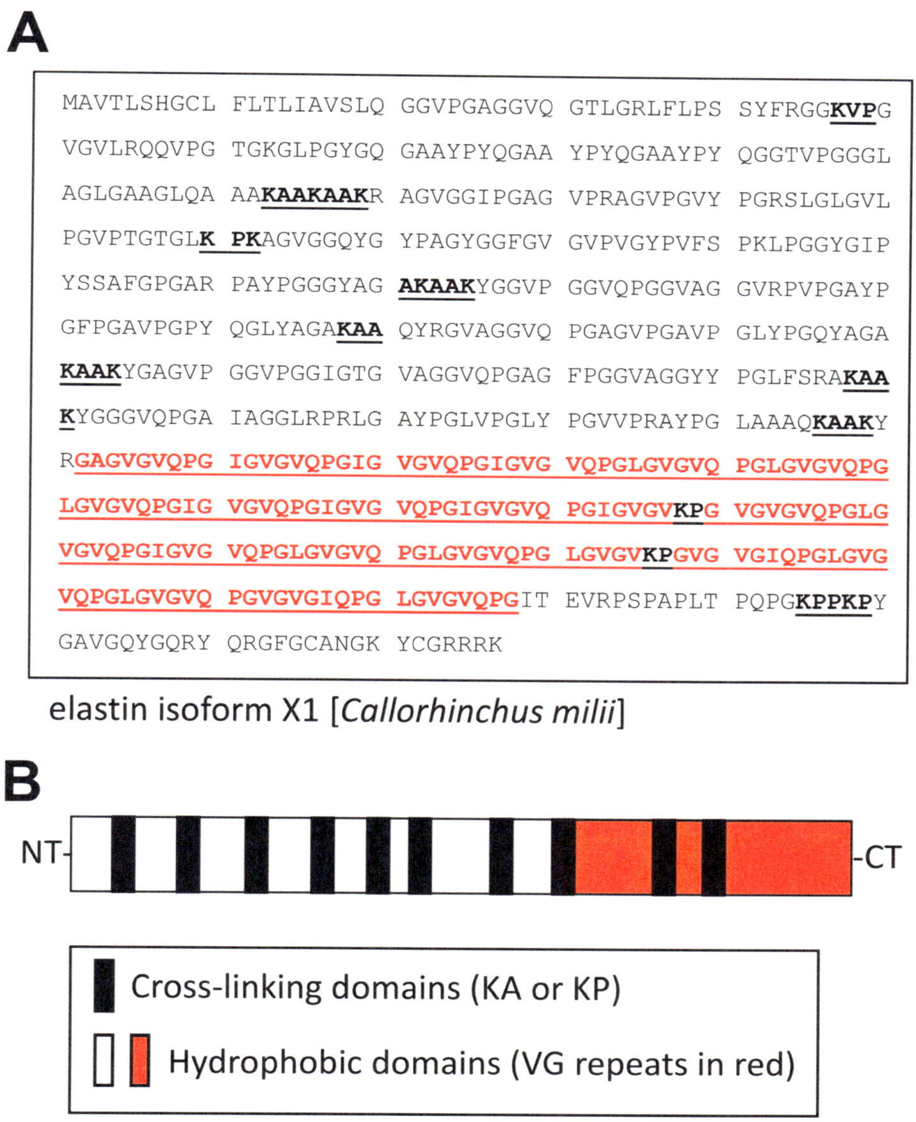

Figure 1.
Primary structure of elephant shark's tropoelastin (Callorhinchus milii, *predicted elastin isoform X1*). *(A) Sequence indicates the position of KA- or KP-crosslinking domains (underlined in black bold). Residues outside the KA/KP domains are enriched in glycine and proline and represent the hydrophobic domains, including those containing the valine/glycine/proline-rich repeats (underlined in red bold). (B) Schematic diagram of the domain structure tropoelastin showing the relative position of cross-linking and hydrophobic domains.*

a single gene. The case of zebrafish is particularly interesting as specialization of *elnb* contributed to the smooth muscle-like characteristics of the bulbus arteriosus, a chamber of the heart zebrafish that is homologous to the aortic trunk of higher vertebrates [13].

Search and identification of elastin sequences in genome databases from different organisms is crucial to delineate its evolutionary history, and to this aim, a significant number of sequences are known today [2, 14, 15]. Nevertheless, an accurate phylogenetic reconstruction of elastin evolution is still quite incomplete. Focus has been placed on different parts of the gene, including a central conserved region, the C-terminal, the 3′-untranslated region and a region presumably resulting from exon replication. However, reaching a unified picture has been difficult. Being an intrinsically disordered protein (IDP) does not make things easier as IDPs lack strict structural constraints, and therefore are more permissive to substitutions [16]. With the only restriction that the conformational flexibility not be altered, IDPs evolve faster than well-folded proteins adding higher complexity to their phylogenetic analyses [17]. To this respect, the soluble monomer of lamprin, the non-collagen/non-elastin major connective tissue component of the lamprey annular cartilage, contains tandem repeats of the sequence GGLGY that are recognized by anti-elastin antibodies targeting the VPG repeats of elastin in a remarkable example of evolutionary convergence [18]. In fact, more distant polypeptides such as some insect proteins or spider silks have also acquired these repeats [19].

Despite the difficulties, phylogenetic trees such as that shown in **Figure 2** based on the central conserved region have been generated.

4. Hypotheses on the evolutionary origin of elastin

As mentioned above, genomic roots of tropoelastin trace back to the elephant shark and related species. Recent publication and open access to whole genome sequences and assemblies of lower vertebrates/chordates have not shown any traceable sign of tropoelastin-related sequences, and that was also true for genomes of invertebrates. These findings (or the lack of them) raise questions as to the origin of tropoelastin and the existence of an ancestral protein. Here, two main hypotheses are proposed to explain its emergence and further evolution: (1) tropoelastin appeared *de novo*; and (2) like other ECM components, it emerged from the assembly of preexisting proteins that eventually gained novel capabilities. What follows discusses evidences and arguments for and against these hypotheses.

4.1 Tropoelastin as a *de novo* protein

Following Darwin's postulates, the general assumption is that new genes evolve from existing ones in an endless, slow-paced journey since the beginning of life. However, recent studies are showing that this has not been always the case and that new genes can arise from the dark depths of the non-coding genome [20]. By gaining the capability of being transcribed and translated, stretches of "junk" DNA can give rise to *de novo* protein products. Interestingly, when deeply studied, *de novo* genes produce firstly dysfunctional or disordered proteins and in many cases with repetitive sequences [21]. Therefore, it is not unreasonable to consider that an ancestral tropoelastin might have emerged as a *de novo* gene. The identification of *de novo* genes is mostly based on the comparison of syntenic regions. This type of analysis has revealed, for instance, that the gene FLJ33706, overexpressed in Alzheimer's disease, appeared in human after the divergence from chimpanzee [22]. Another three human genes of unknown function have been described to originate from chimp non-coding DNA [23]. Unfortunately, synteny of genetic loci is often lost over long evolutionary timescales. Therefore, distant genomic events resulting in *de novo* products are difficult if not impossible to identify. Its emergence in the crossroad of jawed vertebrates places tropoelastin in an unfavorable scenario. Studies so far performed have not found evidences for its *de novo* origin.

4.2 Reorganization or assembly from pre-existing components

It is a recurring theme in the evolutionary history of ECM proteins that the gradual appearance of specific gene families and domains, often in pre-metazoan lineages, allowed thereafter their assembly and formation of matrix components genuine to animals [24]. This has been the case for matrix proteins such as fibrillar or basement membrane collagens, and for matrix-remodeling enzymes like LOX (see **Figure 2**). The late emergence of tropoelastin does not fit with this behavior. As mentioned above, no single tropoelastin-related sequence has been found in genomes back to the elephant shark in the evolutionary scale. Or yes? Before the

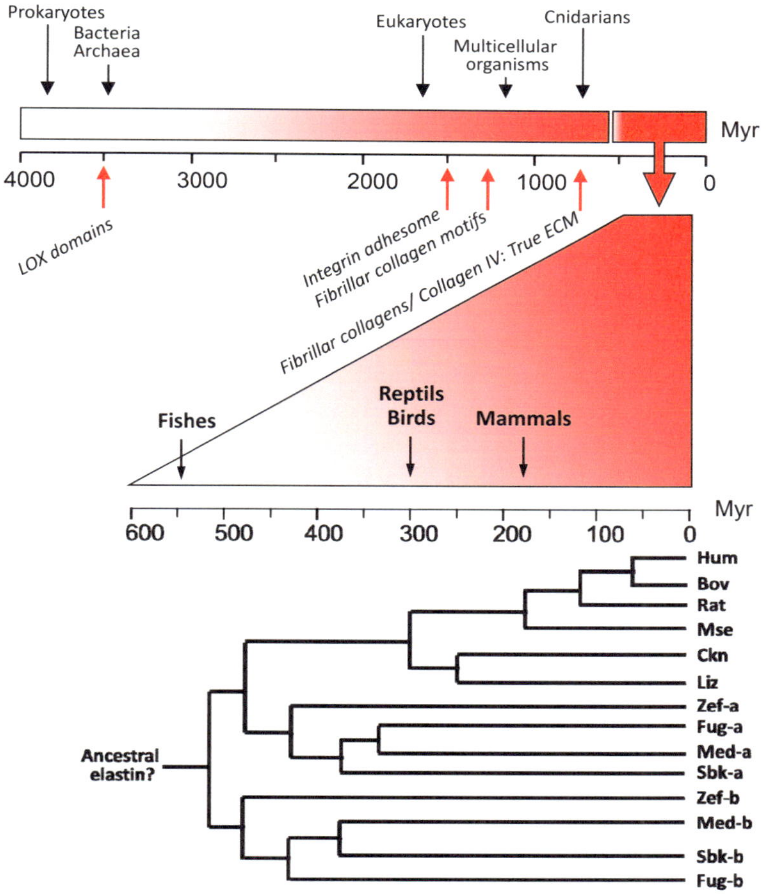

Figure 2.
Phylogenetic tree of the central conserved regions of tropoelastin and chronogram showing divergence times for representatives species. The tree, adapted from Keeley [2], represents generally accepted phylogenetic relationships and was deliberated rooted to propose the existence of a putative ancestral elastin [2]. Chronogram indicates the emergence of relevant phyla (black arrows) as well as that of different ECM or ECM-related motifs or proteins (red arrows) in million of years (Myr). Sequences are as follows: human (Hum), bovine (Bov), rat (Rat), mouse (Mse), chicken (Ckn), lizard (Liz), zebrafish (Zef), japanese pufferfish (Fug), medaka fish (Med), and stickleback (Sbk).

onset of tropoelastin, microfibrils were largely responsible for tissue elasticity in many species. Extracellular matrix structures such as the mesoglea from the cnidarian jellyfish or the blood vessels in invertebrates are functionally elastic due to microfibrils [25, 26]. These supramolecular structures, visualized as beaded filaments under electron microscopy, contain numerous proteins, being fibrillins the major constituent. Like many other ECM components, fibrillins, from which three isoforms exist in humans, fibrillin-1, −2 and − 3, are multidomain proteins that expand along a large polypeptide sequence of almost 3000 aminoacids [27]. The large size and the variety of domains explain the existence of multiple diseases caused by defects in fibrillins, named fibrillinopathies, including various forms of Marfan syndrome, isolated ectopia lentis, kyphoscoliosis, Shprintzen-Golberg syndrome, and stiff skin syndrome, among others [28]. Epidermal growth factor-like domains (EGF and calcium binding EGF) dominate the structure, with 46–47 repeats, followed by transforming growth factor (TGF)-β binding protein domains (TB) and hybrid domains with 8 and 2 repeats, respectively (**Figure 3**). TB domains are shared with latent TGF-β binding proteins (LTBP) and have served to compute phylogenetic reconstructions for these proteins [30]. Using this approach, a TB domain-containing protein was identified in cnidarians, dating the emergence of an ancestral fibrillin to 600 Myr ago. This ancestral fibrillin, not only present in cnidarians, but also in molluscs, annelids, arthropods, echinoderms, urochordates, cephalochordates and lower vertebrates, such as the lamprey, underwent a duplication event at the divergence of jawed and jawless fishes giving to fibrillin-1 and an ancestral fibrillin-2/3. Interestingly, just before this branching, the ancestral fibrillin gained (or reshaped) a domain characterized by a high content of proline- and/or glycine termed "unique region", claimed to provide a flexible behavior and for which a specific function has not yet been demonstrated (see also **Figure 3**) [30]. In fact, when looking carefully to the these domains from different species including jawed and jawless fishes, a clear evolution from a short sequence with just a few proline and glycine residues as seen in the ascidian *Ciona intestinalis* or the lancelet *Branchiostoma floridae* to a longer segment that increases its proline/glycine content

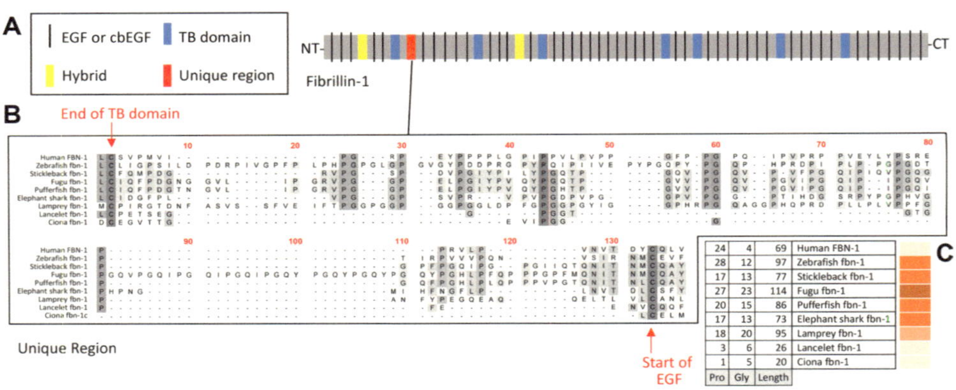

Figure 3.
*Domain structure of fibrillin-1 and sequence alignment of the unique region from different species. (A) Overview of the domain structure of fibrillin-1 showing the relative positions of epidermal growth factor (EGF) and calcium-binding EGF domains (cbEGF), transforming growth factor-β binding protein domains (TB), hybrid domains and the unique region. (B) A group of unique regions from fibrillins representing different species relevant for the emergence of tropoelastin were aligned using the ClustalW algorithm [29]. Aligned segments are flanked by conserved cysteines from the end of the first TB domain (left) and the start of the EGF (or cbEGF-like) domains (right) as indicated. (C) Table shows the number of prolines (Pro) and glycines (Gly) within the unique region of fibrillin in the analyzed species as well as their total length of residues. Bar at the right of the table assigns a color intensity to these unique regions based on the number of proline/glycine residues as it is used in the schematical diagram of **Figure 4**.*

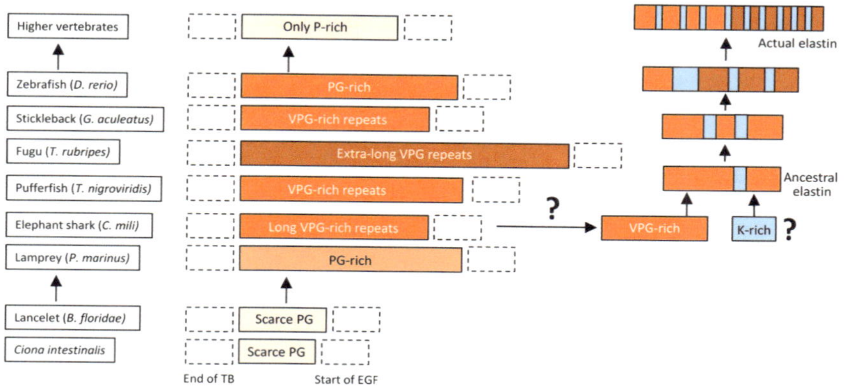

Figure 4.
*Hypothesis for the emergence of elastin from the unique region of fibrillins from fish species. Schematical diagram proposing the evolutionary origin of elastin from sequences of the unique region of fibrillins in the fish lineage. Using a color code based on PG content (**Figure 3**), the figure shows that unique regions in the invertebrates* Ciona intestinalis *and* Branchiostoma floridae, *being short and containing scarce PG residues, progressively increases in size and PG content, displaying a significant number of VPG-repeats from elephant shark onwards. One of such domains in an ancient relative of the elephant shark might have contributed to the assembly of a primordial elastin. This ancestral protein could evolve further by gaining K-rich cross-linking domains and undergoing extensive exon expansion.*

in the lamprey and progresses to extra-long fragments with a significant number of VPG-containing repeats, such as the Japanese pufferfish *Takifugu rubripes*. These sequences remarkably resembles the hydrophobic domains of tropoelastin, particularly those seen in the first jawed fish such as the elephant shark (see **Figure 1**), and, intriguingly, their appearance is evolutionary coincident with that of elastin and the high blood pressure closed circulatory systems in these organisms. It is tempting to speculate that the VPG-containing repeats from fibrillin-1 contributed to the assembly of an ancestral elastin (**Figure 4**). Subsequent changes giving to KA or KP domains or their incorporation from an unknown ancestor, as well as extensive domain duplication and expansion, might have ended up sculpting the elastin backbone as it is found in elastin-expressing species living today. In fact, in these organisms, microfibrils provide the scaffolding platform where elastogenesis takes places, making the entire microfibril-elastic fiber unit the material responsible for the biomechanical properties of tissues such as the lung, blood vessels and skin [31]. Within this context, it has been speculated that the unique region in fibrillin-1 evolved to support the interaction with elastin [30, 32]. Considering that intrinsically disorder regions can use their flexibility to allow the association between two (or more) IDPs, it is not misconceived to think that the acquisition of this domain by an ancestral tropoelastin may have served to mutually establish the binding both fibrillin-1 and elastin [33]. Constraints to keep the structural conformation rather than the primary sequence may have then blurred the phylogenetic relationships, making difficult to trace back the origin of this genomic event. Curiously, the unique region of fibrillin-1 in higher vertebrates dynamically evolved losing the VPG-containing repeats while still keeping a high content of proline residues, perhaps reflecting novel requirements in the fibrillin-1/elastin interaction during elastogenesis in these species.

5. Concluding remarks

Whether elastin evolved as a *de novo* protein or derived from a pre-existing fibrillin-1 (or any other unknown) gene remains with the available genomic information as an obscure enigma. The intention of this chapter was to bring together the current knowledge about the evolutionary history of elastin and to discuss the hypotheses that eventually may explain its origin. While this is certainly in the realm of speculation, it is hoped that the sequencing and annotation of more genomes as well as the advent of further molecular and genomic analyses will permit to get more insight about the evolutionary roots of this fascinating protein.

6. Methods

Sequences used in this work are:
Elephant shark predicted elastin isoform X1 (*Callorhinchus milii*) [Genbank XP_007894595].
Human fibrillin-1 preproprotein (*Homo sapiens*) [Genbank NP_000129].
Zebrafish fibrillin-1 (*Danio rerio*) [Genbank XP_017207479].
Stickleback fibrillin-1 (*Gasterosteus aculeatus*) [UniProtKB G3PX14].
Fugu fibrillin-1 isoform X1 (*Takifugu rubripes*) [Genbank XP_003969883.1].
Pufferfish fibrillin 1 (*Tetraodon nigroviridis*) [UniProtKB H3C692].
Elephant shark predicted fibrillin-1-like, partial (*Callorhinchus milii*) [Genbank XP_007909428].
Sea lamprey putative fibrillin-1 (*Petromyzon marinus*) [UniProtKB S4RBV9].
Lancelet putative fibrillin-1 (*Branchiostoma floridae*) [Genbank XP_002601550].
Fibrillin-1 (*Ciona intestinalis*) [Genbank XP_009858101].

Acknowledgements

We thank M. Mar Alba (Universitat Pompeu Fabra, Barcelona, Spain) for helpful comments.

I acknowledge support of the publication fee by the CSIC Open Access Publication Support Initiative through its Unit of Information Resources for Research (URICI).

Author details

Fernando Rodriguez-Pascual
Centro de Biología Molecular "Severo Ochoa" Consejo Superior de Investigaciones Científicas (C.S.I.C.), Universidad Autónoma de Madrid (Madrid), Madrid, Spain

Address all correspondence to: frodriguez@cbm.csic.es

References

[1] Kozel, B. A., and Mecham, R. P. (2019) Elastic fiber ultrastructure and assembly. Matrix Biology **84**, 31-40

[2] Keeley, F. W. (2013) The Evolution of Elastin. in *Evolution of Extracellular Matrix* (Keeley, F. W., and Mecham, R. P. eds.), Springer Berlin Heidelberg, Berlin, Heidelberg. pp 73-119

[3] Engel, J., and Chiquet, M. (2011) An Overview of Extracellular Matrix Structure and Function. in *The Extracellular Matrix: an Overview* (Mecham, R. P. ed.), Springer Berlin Heidelberg, Berlin, Heidelberg. pp 1-39

[4] Tarakanova, A., Yeo, G. C., Baldock, C., Weiss, A. S., and Buehler, M. J. (2018) Molecular model of human tropoelastin and implications of associated mutations. Proceedings of the National Academy of Sciences **115**, 7338-7343

[5] Okuyama, K. (2008) Revisiting the molecular structure of collagen. Connective tissue research **49**, 299-310

[6] Muiznieks, L. D. M. D., Weiss, A. S. W. S., and Keeley, F. W. K. W. (2010) Structural disorder and dynamics of elastinThis paper is one of a selection of papers published in this special issue entitled "Canadian Society of Biochemistry, Molecular & Cellular Biology 52nd Annual Meeting — Protein Folding: Principles and Diseases" and has undergone the Journal's usual peer review process. Biochemistry and Cell Biology **88**, 239-250

[7] Roberts, S., Dzuricky, M., and Chilkoti, A. (2015) Elastin-like polypeptides as models of intrinsically disordered proteins. FEBS Lett **589**, 2477-2486

[8] Rauscher, S., and Pomès, R. (2012) Structural Disorder and Protein Elasticity. in *Fuzziness: Structural Disorder in Protein Complexes* (Fuxreiter, M., and Tompa, P. eds.), Springer US, New York, NY. pp 159-183

[9] Rauscher, S., Baud, S., Miao, M., Keeley, Fred W., and Pomès, R. (2006) Proline and Glycine Control Protein Self-Organization into Elastomeric or Amyloid Fibrils. Structure **14**, 1667-1676

[10] Schräder, C. U., Heinz, A., Majovsky, P., Karaman Mayack, B., Brinckmann, J., Sippl, W., and Schmelzer, C. E. H. (2018) Elastin is heterogeneously cross-linked. J Biol Chem **293**, 15107-15119

[11] Rodriguez-Pascual, F., and Slatter, D. A. (2016) Collagen cross-linking: insights on the evolution of metazoan extracellular matrix. Scientific Reports **6**, 37374

[12] Monahan-Earley, R., Dvorak, A. M., and Aird, W. C. (2013) Evolutionary origins of the blood vascular system and endothelium. J Thromb Haemost **11 Suppl 1**, 46-66

[13] Moriyama, Y., Ito, F., Takeda, H., Yano, T., Okabe, M., Kuraku, S., Keeley, F. W., and Koshiba-Takeuchi, K. (2016) Evolution of the fish heart by sub/neofunctionalization of an elastin gene. Nature Communications **7**, 10397

[14] Chung, M. I. S., Ming, M., Stahl, R. J., Chan, E., Parkinson, J., and Keeley, F. W. (2006) Sequences and domain structures of mammalian, avian, amphibian and teleost tropoelastins: Clues to the evolutionary history of elastins. Matrix Biology **25**, 492-504

[15] He, D., Chung, M., Chan, E., Alleyne, T., Ha, K. C. H., Miao, M., Stahl, R. J., Keeley, F. W., and Parkinson, J. (2007) Comparative genomics of elastin: Sequence analysis

of a highly repetitive protein. Matrix Biology **26**, 524-540

[16] Piontkivska, H., Zhang, Y., Green, E. D., Program, N. C. S., and Elnitski, L. (2004) Multi-species sequence comparison reveals dynamic evolution of the elastin gene that has involved purifying selection and lineage-specific insertions/deletions. BMC Genomics **5**, 31-31

[17] Brown, C. J., Johnson, A. K., Dunker, A. K., and Daughdrill, G. W. (2011) Evolution and disorder. Current Opinion in Structural Biology **21**, 441-446

[18] Robson, P., Wright, G. M., Sitarz, E., Maiti, A., Rawat, M., Youson, J. H., and Keeley, F. W. (1993) Characterization of lamprin, an unusual matrix protein from lamprey cartilage. Implications for evolution, structure, and assembly of elastin and other fibrillar proteins. Journal of Biological Chemistry **268**, 1440-1447

[19] Xu, M., and Lewis, R. V. (1990) Structure of a protein superfiber: spider dragline silk. Proceedings of the National Academy of Sciences **87**, 7120-7124

[20] Levy, A. (2019) How evolution builds genes from scratch. Nature **574**, 314-316

[21] Toll-Riera, M., Radó-Trilla, N., Martys, F., and Albà, M. M. (2011) Role of Low-Complexity Sequences in the Formation of Novel Protein Coding Sequences. Molecular Biology and Evolution **29**, 883-886

[22] Li, C.-Y., Zhang, Y., Wang, Z., Zhang, Y., Cao, C., Zhang, P.-W., Lu, S.-J., Li, X.-M., Yu, Q., Zheng, X., Du, Q., Uhl, G. R., Liu, Q.-R., and Wei, L. (2010) A Human-Specific De Novo Protein-Coding Gene Associated with Human Brain Functions. PLOS Computational Biology **6**, e1000734

[23] Knowles, D. G., and McLysaght, A. (2009) Recent de novo origin of human protein-coding genes. Genome Research **19**, 1752-1759

[24] Rodríguez-Pascual, F. (2019) How evolution made the matrix punch at the multicellularity party. Journal of Biological Chemistry **294**, 770-771

[25] Reber-Müller, S., Spissinger, T., Schuchert, P., Spring, J., and Schmid, V. (1995) An Extracellular Matrix Protein of Jellyfish Homologous to Mammalian Fibrillins Forms Different Fibrils Depending on the Life Stage of the Animal. Developmental Biology **169**, 662-672

[26] McConnell, C. J., DeMont, M. E., and Wright, G. M. (1997) Microfibrils provide non-linear elastic behaviour in the abdominal artery of the lobster Homarus americanus. J Physiol **499 (Pt 2)**, 513-526

[27] Hubmacher, D., and Reinhardt, D. P. (2011) Microfibrils and Fibrillin. in *The Extracellular Matrix: an Overview* (Mecham, R. P. ed.), Springer Berlin Heidelberg, Berlin, Heidelberg. pp 233-265

[28] Aubart, M., Benarroch, L., Arnaud, P., Collod-Béroud, G., and Jondeau, G. (2016) Molecular Genetics of the Fibrillinopathies. *eLS (Wiley Online* Library*)*, 1-13

[29] Thompson, J. D., Higgins, D. G., and Gibson, T. J. (1994) CLUSTAL W: improving the sensitivity of progressive multiple sequence alignment through sequence weighting, position-specific gap penalties and weight matrix choice. Nucleic acids research **22**, 4673-4680

[30] Piha-Gossack, A., Sossin, W., and Reinhardt, D. P. (2012) The evolution of extracellular fibrillins and their functional domains. PLoS One **7**, e33560-e33560

[31] Faury, G. (2001) Function–structure relationship of elastic arteries in evolution: from microfibrils to elastin and elastic fibres. Pathologie Biologie **49**, 310-325

[32] Rock, M. J., Cain, S. A., Freeman, L. J., Morgan, A., Mellody, K., Marson, A., Shuttleworth, C. A., Weiss, A. S., and Kielty, C. M. (2004) Molecular Basis of Elastic Fiber Formation: CRITICAL INTERACTIONS AND A TROPOELASTIN-FIBRILLIN-1 CROSS-LINK. Journal of Biological Chemistry **279**, 23748-23758

[33] Oldfield, C. J., and Dunker, A. K. (2014) Intrinsically Disordered Proteins and Intrinsically Disordered Protein Regions. Annual Review of Biochemistry **83**, 553-584

Chapter 5

The Interplay of ECM-Based Graft Materials and Mechanisms of Tissue Remodeling

Jason P. Hodde and Michael C. Hiles

Abstract

Wound healing is a complex natural process that involves the recruitment of cells, the renewal of tissue composition, and the reinforcement of structural tissue architecture. Following ischemic injury or chronic disease, wound healing is delayed, and can often result in chronic inflammation or permanent morbidity. Tissue engineering strategies to harness the wound healing process include the use of naturally derived extracellular matrix (ECM) scaffolds with inherent bioactivity to both passively facilitate and actively direct healing toward a successful resolution. As the body heals, the properly designed ECM scaffold is gradually remodeled and integrated into the body, leaving behind organized tissue that provides long-term strength. Herein we explain the interplay of the ECM (i.e., its complex composition and bioactivity) with the cells of the body throughout the process of tissue remodeling, thus explaining how even a tissue-engineered xenograft material can direct the body to restore itself.

Keywords: wound healing, extracellular matrix, bioactivity, tissue remodeling, xenograft

1. Introduction: extracellular matrix as an implantable graft material

Biologic materials used to repair soft tissue defects must be strong and easy to handle during implantation, but they must also be able to support tissue integration and maturation once implanted. ECM-based biologic grafts have been widely used in surgery over the last two decades. They are a good choice for surgeons because they can be safely implanted in contaminated settings where synthetic materials are contraindicated. Even though synthetic mesh materials continue to be favored in general surgical practice because of their versatility and low cost, they remain susceptible to chemical degradation over time, can create physical tissue erosion due to mismatches in their mechanical properties with the surrounding tissues, and may undergo encapsulation following placement because the body views them as foreign materials [1]. Of critical importance in many applications, synthetics can provide a nidus for microorganism growth; therefore, if they become infected when in the body, they typically need to be removed [2].

ECM biomaterials derived from natural tissue sources, however, have generally provided adequate strength, resistance to infection, and stability over time such that they make adequate materials for soft tissue reconstruction [3].

These materials can be obtained as autografts or allografts, but autografts result in donor site morbidity, while cadaveric allograft tissues may transmit disease, are inherently inconsistent, and are typically quite expensive.

Recent years have seen the advent of multiple off-the shelf tissue-based ECM biomaterials that claim to provide an optimal healing environment for soft tissues. They can be obtained from a wide variety of mammalian tissues, processed using a wide range of chemicals and cross-linking agents, or can be provided in such a way that retains the information-rich scaffold into which adjacent cells migrate to create a replacement tissue (**Table 1**). Many studies have shown constructive, functional tissue remodeling with partial restoration of site-appropriate tissue using these graft materials [4–7], yet this is not always the case. Less favorable outcomes include the accumulation of serous fluid at the implant site, rapid degradation of the graft material with associated mechanical failure, or a lack of biomaterial integration with the patient's tissues, resulting in a foreign body response [8, 9]. These less-than-favorable outcomes typically have been associated with variations in manufacturing methods that result in the failure of the material to maintain nature's natural composition and three-dimensional architecture that makes the extracellular matrix (ECM) the ideal template for tissue repair and regeneration.

Materials that are minimally processed most closely recapitulate the structure and function of the original tissue while providing a safe, biocompatible material for soft tissue reconstruction. The natural ECM, when retained in its complex arrangement of matrix proteins and associated factors, can provide the key extracellular signals and inherent bioactivity needed to restore damaged tissues to their natural state [7]. This complexity allows the naturally occurring biologic graft to completely integrate with the recipient's tissues and cells to ultimately form a vascularized, highly organized tissue structure that resembles the native tissue structure and architecture [4, 7, 10, 11].

Product	Source	Crosslinking agent	Sterilization
Alloderm	Human dermis	N/A	N/A
AlloMax	Human dermis	N/A	Gamma radiation
Biodesign	Porcine small intestine	N/A	EtO
Gentrix	Porcine urinary bladder	N/A	E-beam
GraftJacket	Human dermis	N/A	N/A
Meso BioMatrix	Porcine mesothelium	N/A	EtO
MicroMatrix	Porcine urinary bladder	N/A	E-beam
Miroderm	Porcine liver	N/A	E-beam
OASIS	Porcine small intestine	N/A	EtO
Peri-Guard	Bovine pericardium	Glutaraldehyde	Liquid chemical
Permacol	Porcine dermis	HMDI	Gamma radiation
Strattice	Porcine dermis	N/A	E-beam
Tutoplast	Human pericardium	N/A	Gamma radiation
XenMatrix	Porcine dermis	N/A	E-beam

EtO ethylene oxide, E-beam Electron beam irradiation, HMDI hexamethylene diisocyanate.

Table 1.
Source tissue and post-decellularization processing steps of some common commercially available ECM biomaterials.

2. Extracellular matrix as bioactive structure

The ECM is a three-dimensional network of extracellular macromolecules, such as collagens, glycoproteins, proteoglycans, and glycosaminoglycans, that provides structural and biochemical support to surrounding cells. Because of different structural and mechanical requirements, the composition of ECM varies from tissue to tissue; however, providing a structure for cell adhesion, directing cell-to-cell communication, and regulating cell processes such as growth, migration and differentiation are common functions of the ECM [12].

Regardless of the source, ECM is a complex three-dimensional scaffold consisting of structural and functional proteins and components arranged in a tissue-specific orientation [12]. The ECM components directly interact with fibroblasts, endothelial cells, and macrophages to maintain a natural and functional homeostatic environment through a process known as dynamic reciprocity (**Figure 1**) [13]. When injury occurs and the natural equilibrium is disrupted, the dynamic environment that exists between the ECM and cells orchestrates acute inflammation, wound healing and tissue remodeling to regain function and restore homeostasis. After injury occurs and the ECM is damaged, a biologic graft can be implanted to provide a surrogate matrix structure that allows dynamic reciprocity to begin immediately, ultimately achieving tissue restoration via the process of constructive tissue remodeling.

Endogenous ECM functions as the intended bioactive structure when normal tissue turnover is taking place or when no significant tissue loss is encountered. The body has a remarkable ability to self-renew, in large part due to the instructional

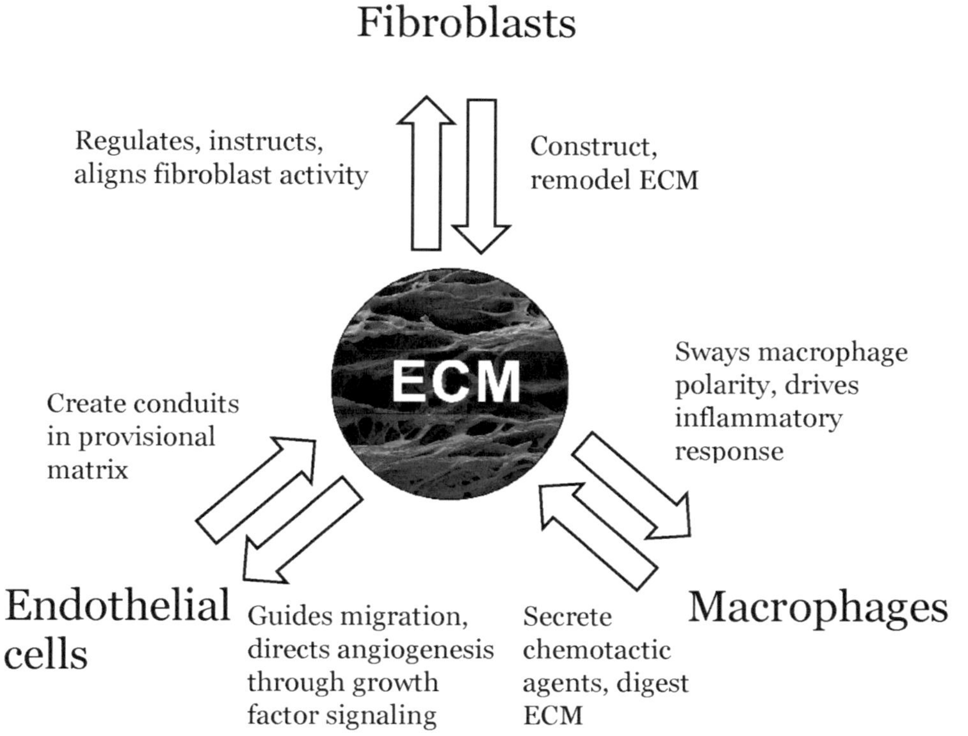

Figure 1.
Examples of dynamic reciprocity of fibroblasts, macrophages, endothelial cells (angiogenesis), and the extracellular matrix (ECM) during wound healing. These interactions occur through signals such as growth factors and/or binding of cells to the ECM.

nature of the ECM, but in the presence of significant tissue loss, large areas of trauma, or surgical reconstructions, there is a need for an exogenous material to augment and to bring order to somewhat chaotic processes. An exogenous ECM can serve as this bioactive, instructive, and even mechanical blueprint for a constructive tissue remodeling process [7, 14, 15].

3. Extracellular matrix and constructive tissue remodeling

Constructive tissue remodeling is more than just another word for wound healing or for tissue repair. The stages of wound healing include initial hemostasis, characterized by clot formation; inflammation, characterized by the deposition of inflammatory and progenitor cells, leading to the formation of granulation tissue; proliferation, where resident cells secrete growth factors and cytokines and collagen deposition occurs; and remodeling, where the newly formed tissue matures and collagen strength increases to meet the demands of the body [16] (**Figure 2**). Tissue repair results in the formation of scar tissue, which is known to be less strong than native tissue and can therefore be more susceptible to reinjury [5].

Unlike the tissue repair process that occurs in the absence of a biologic graft material, the constructive tissue remodeling process that can be directed by an ECM graft leads to a more natural healing process in the recipient that is characterized by the deposition of organized connective tissue, rather than just chaotic scar [17]. The ideal ECM graft is characterized by an open matrix structure to allow for rapid cellular ingrowth. It is also characterized by the presence of structural collagens and non-collagen ECM components (such as messenger nucleic acids, growth factors, glycoproteins, proteoglycans, and glycosaminoglycans), which act to facilitate the renewal of natural dynamic reciprocity [18]. When tissue homeostasis is disrupted, the biologic graft plays the role of the recipient's natural ECM and works to bridge the recipient's cells across the wound to ultimately restore a homeostatic environment. The restoration of homeostasis following injury in the presence of a biologic graft occurs through the constructive process of tissue remodeling.

Tissue remodeling is a process of tissue restoration that improves upon the scar tissue outcome typically achieved by tissue repair. It can be divided into three separate phases: 1) Cell recruitment; 2) Tissue renewal; and 3) Tissue reinforcement.

During cell recruitment, the remodeling process starts when the body's inflammatory and progenitor cells populate the biologic graft and release cytokines and growth factors that bind to the graft and recruit collagen-secreting fibroblasts [18, 19]. In this phase, the graft primarily acts as a scaffold material to support the population of the open ECM structure by the patient's own cells.

As remodeling progresses, the patient's macrophages and fibroblasts in the newly populated matrix work together with matrix-bound signaling factors to renew the tissue through the complementary processes of phagocytosis, collagen deposition, and angiogenesis. In this phase, the biologic graft is gradually replaced by the patient's own tissue and cells [18, 19].

Over the medium to long term, the resident fibroblasts secrete cytokines and growth factors to signal reinforcement of the deposited tissue through the processes of additional collagen deposition and maturation, resulting in a strong, repaired tissue [10, 20–22]. In this phase, the biologic graft is no longer needed as the patient's own collagen has gradually matured into a stable structure that has long-term strength but is entirely the patient's own [20–22]. The resulting tissue structure is mature, organized and strong, and can withstand (and is even driven by) the natural physiological forces that it encounters [17, 23].

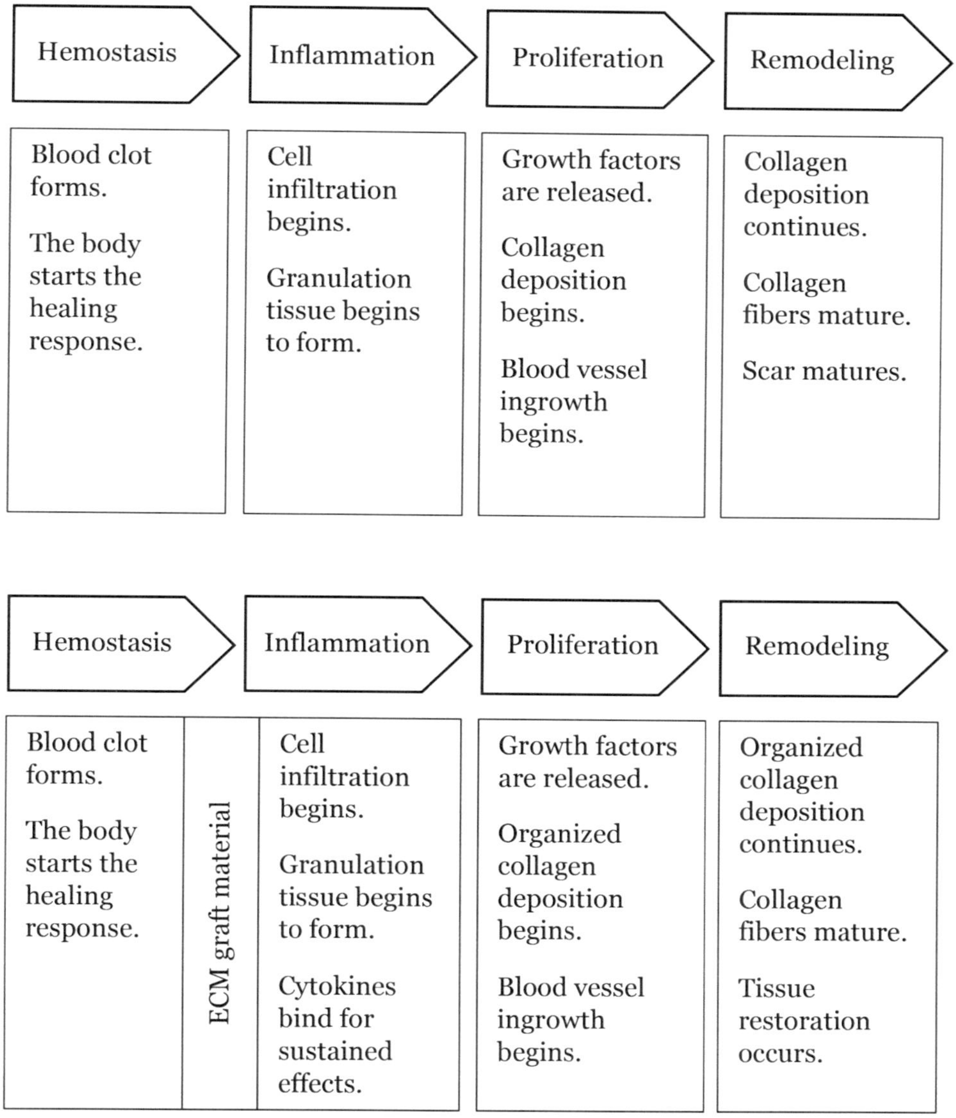

Figure 2.
The phases of wound healing and the processes involved in each stage. The addition of an ECM graft material shortly after the injury occurs results in a more natural wound healing response than in its absence.

A biologic graft with the correct composition and three-dimensional architecture directs the patient's body to replace itself – to completely remodel – rather than to heal through a tissue repair process that results in chaotic, weak, and ineffective scar tissue formation [20–22]. By providing the correct cues to help the body restore itself, the graft provides both an essential temporary structure and the local tissue instructions to lead the patient to achieve a natural repair (**Figure 3**).

4. Mechanisms of action for ECM-directed tissue remodeling

An ECM-based biologic graft that has been optimally processed to harness the tissue remodeling properties of nature acts more than just a mechanical tissue reinforcement device. While mechanical reinforcement is still the primary mechanism of action

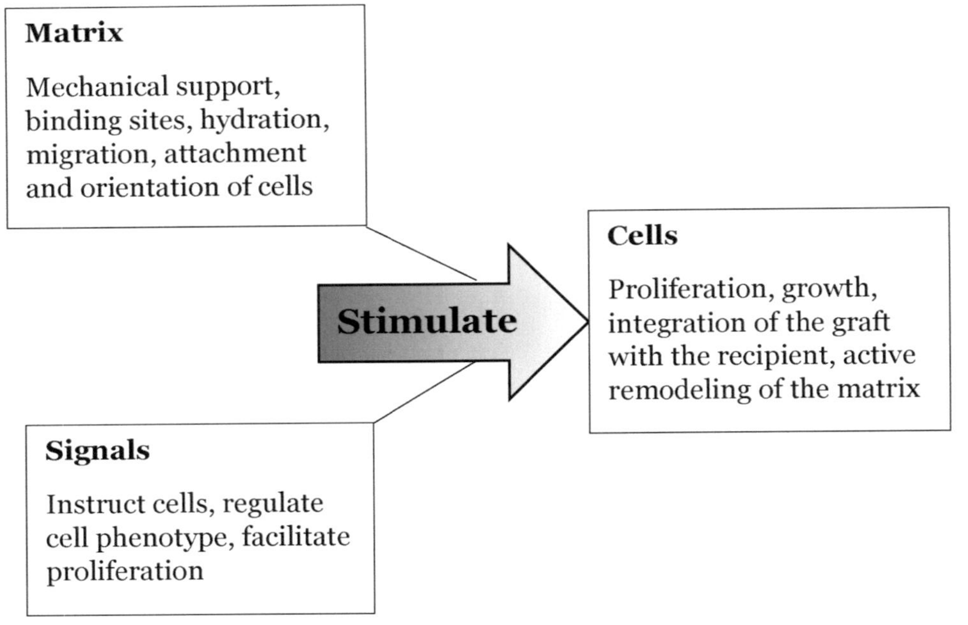

Figure 3.
Mechanisms of action for ECM-directed tissue remodeling. The ECM graft initially provides for a direct mechanical tissue repair that has inherent strength. It also provides a matrix structure for the support, attachment, and orientation of cells. The ECM graft has the ability, through its inherent composition, to modulate the local wound environment to have a direct effect on endogenous growth factors and cytokines. The graft can provide signals of its own, which may include growth factors, binding sequences on extracellular matrix proteins, or other endogenous factors provided by the recipient. Signals control and modify cells and other elements. Together the ECM and signals stimulate cell division, proliferation, growth, and integration of the ECM graft with the recipient.

for these materials, additional mechanisms of action include: providing a porous tissue scaffold matrix structure to allow for fibroblast infiltration and population; altering the surrounding wound environment by modulating local cytokine activity; and, optimally, acting as a reservoir for growth factors and signaling molecules that can be used by the patient as tissue renewal and reinforcement progress (**Table 2**).

4.1 Mechanical reinforcement during surgical repair

Poor wound healing after trauma, surgery, or due to chronic disease is the consequence of a poorly regulated tissue repair response that directly effects the processes of inflammation, angiogenesis, matrix deposition, and cell recruitment [24]. As a result, tissue healing typically takes a significant time to achieve in patients with advanced age or with comorbidities. Prolonged mechanical reinforcement is often needed to get proper approximation of the wound edges and to bolster the anatomy until tissue ingrowth is sufficient to achieve the required strength to maintain tissue integrity. This mechanical reinforcement mechanism is the primary (and often only) means by which most implantable devices achieve their effect. For example, synthetic mesh materials, such as polypropylene or polytetrafluoroethylene, derive their reinforcement benefit from the strength of their fibers at implant but never completely integrate with the patient's tissues over time [25]. Synthetic materials are often recognized as foreign by the body – as a material that needs to be removed or expunged [26]. When this occurs, an inflammatory response is initiated by the patient's immune system, setting up a chronic inflammatory state that never resolves and can result in chronic pain and fibrosis [26].

	Mechanical Surgical Repair	Tissue Scaffold Matrix Structure	Modulation of Endogenous Cytokines	Delivery of Exogenous Cytokines
Synthetic mesh implant	✓	×	×	×
Bio-synthetic mesh implant	✓	×	×	×
Cross-linked ECM biologic graft	✓	✓/×	×	×
Purified ECM biologic graft	✓	✓	×	×
Naturally complex ECM biologic graft	✓	✓	✓	✓

Table 2.
Mechanisms of action for different types of implantable graft materials. While all implantable materials serve a mechanical function to reinforce soft tissue, synthetic and biosynthetic materials fail to provide a matrix structure and complex composition that is designed to positively interact with the wound healing environment and lead to constructive tissue remodeling that is seen with naturally complex ECM biologic graft materials.

For a well-designed biologic ECM graft, the mechanical means of tissue support remains its primary mechanism of action. The ECM graft must allow the passage of suture and reinforce the area of weakness under significant pull-out force. It must also provide tensile strength and mechanical compliance commensurate with the surrounding tissues. Unlike synthetic or even many biosynthetic materials, such as poly-4-hydroxybutyrate (P4HB), ECM-based biologic devices are not meant to be static implants but are designed to fully integrate with the patient over time. Their mechanical properties change after implant as they undergo interaction with the patient's cells, tissues, and the local wound environment [27] and must therefore be designed to retain their mechanical integrity even while actively participating in the process of tissue renewal. The dynamic process of tissue remodeling is a balance of ECM graft degradation with the formation of new patient-derived collagen, meaning that an ECM graft must be designed with known strength requirements and degradation rates to keep the repair intact during all phases of tissue remodeling: 1) Cell recruitment; 2) Tissue renewal; and 3) Tissue reinforcement (**Figure 4**) [20, 28].

4.2 Providing a tissue scaffold matrix structure

When foreign materials are implanted into the body, they are quickly recognized by the immune system as something either to rapidly destroy or to compartmentalize [29]. The body accomplishes these activities by secreting inflammatory enzymes and pH modifiers or by recruiting an army of macrophages to form a scarified wall around the implant. While permanent synthetic materials and crosslinked biologic grafts are typically walled off by the recipient because they are resistant to degradation [30], biosynthetic matrices are often hydrolyzed or otherwise degraded over time without allowing complete tissue integration and permanent reinforcement to occur [31].

Purified biologic ECM grafts typically contain few of the naturally occurring macromolecules of the complex ECM because they have been deconstructed with chemicals and then "purified" into single-component constructs or reconstituted into single-component implants. While this type of graft material can still act as a matrix structure to support cell ingrowth, the lack of complex signaling macromolecules from the natural ECM and its susceptibility to matrix-degrading enzymes, such as collagenases, limits its ability to actively promote fibroblast and endothelial cell proliferation and secretion of new ECM [32, 33].

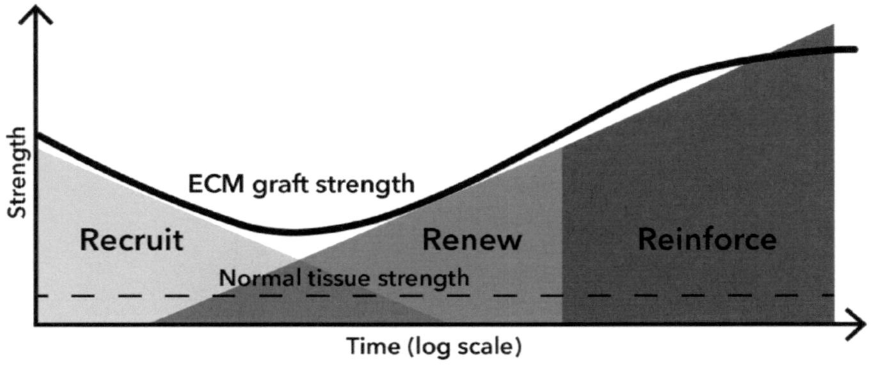

Figure 4.
ECM-based graft materials must be designed to withstand physiologic forces while undergoing the active processes of tissue remodeling and tissue integration following implant. The overall repair strength must be maintained well above the normal tissue strength required to keep the repair intact while facilitating cell recruitment, tissue renewal, and tissue reinforcement.

Non-crosslinked biologic ECM grafts that have been processed to retain the composition and architecture of healthy ECM are neither encapsulated nor degraded upon implant [7]. Instead, they contain the complex information of the natural ECM that makes them an ideal scaffold environment upon which cells can move and proliferate, allowing for colonization of fibroblasts and endothelial cells, the eventual secretion of growth factors, and the deposition of a collagen matrix [10]. The porous nature of the ECM scaffold provides not only the structure and interstices for ingrowth but also the recognition and binding sites that facilitate cellular attachment and migration [10]. During the process of tissue renewal, the porous matrix structure of the non-crosslinked ECM graft allows for angiogenesis and ultimately the removal of byproducts of cellular metabolism, facilitating the process of tissue remodeling that is essential to obtaining a long-lasting, strong, and permanent repair [10, 34].

4.3 Modulating endogenous cytokine activity

The local wound environment is characterized by a dynamic milieu of signaling factors designed to shepherd an injury through the four phases of wound healing and to ultimately restore tissue strength and homeostasis [16]. In most instances this occurs in a well-defined series of events leading to complete tissue restoration that is modulated directly by the local ECM. Because the ECM is laden with macromolecules that explicitly bind cytokines and alter their half-lives, bioactivities, and concentrations, the presence of a healthy ECM in the local wound environment is essential for tissue remodeling to occur. When the ECM is corrupt, it cannot support tissue restoration and chronic inflammation results [35].

Chronic, non-healing wounds are characterized by increased levels of pro-inflammatory cytokines, increased levels of MMPs, and low levels of growth factors known to stimulate wound closure [36, 37]. They are highly inflamed and proteolytic, have become stalled in the inflammation stage of wound healing, and cannot support fibroblast function [38]. In cases such as this, replacing the damaged ECM with a healthy ECM-based biologic graft can alter the local wound environment by modulating the endogenous cytokine profile of the injured area and stimulating normal fibroblast and endothelial cell function [39].

This tertiary mechanism of action for ECM-directed tissue remodeling, endogenous cytokine modulation, harnesses the natural structure and composition of the ECM to direct tissue remodeling down a productive pathway [37]. Unlike synthetic and biosynthetic materials that contain no ECM-binding sites and cannot directly influence the composition of the natural wound environment; unlike crosslinked ECM biologic graft materials which have had their binding sites obscured by the crosslinking process; and unlike purified biologic ECM grafts that are limited in the types of cytokines that can interact with them; well-designed, non-crosslinked, biologic ECM graft materials have been shown to positively alter the local environment and lead to constructive tissue remodeling and wound healing [10, 37, 39, 40].

4.4 Acting as a cytokine reservoir

Matrix biologists have long regarded the ECM as a repository for latent bioactivity in the form of growth factors, cytokines, and more recently, messenger nucleic acid depots. Even in their dehydrated state, these factors retain their potency and structure because they are tightly bound to proteins that protect them from degradation [41, 42]. Also, recently, science has uncovered the remarkable ability for these embedded matrix molecules to modulate cellular activity across species and after long periods of dormancy. Porcine growth factors can activate human cells, and vice-versa, with predictable potency and expected effects, even after dehydration and sterilization [41, 42]. It is this growth factor and cytokine repository that separates a complex biologic ECM graft from other types of non-instructional implant materials.

After implantation, a complex biologic ECM graft plays the role of the innate ECM, interacting with the patient's cells through dynamic reciprocity to direct tissue repair down a positive, active state of wound healing and toward an organized repair that resembles native tissue structure and architecture rather than scar tissue. When its role has been fully realized, an ECM graft becomes completely replaced by patient tissue and removed from the body through the normal process of matrix turnover, leaving no graft components behind [43]. In many ways it is the repository of latent bioactivity that allows the well-designed ECM graft to stimulate transformation of itself, by the patient's cells, into a new, complex and complete, functional tissue.

5. Summary: extracellular matrix past, present, and future

ECM graft materials have been used surgically for decades, but historically they have been enzymatically stripped of their biological information, chemically cross-linked to enhance their durability (while quite effectively silencing their biological activity), or otherwise adulterated in such a way as to act much more like synthetic mesh than a truly instructive matrix [34, 44]. A more modern approach to ECM graft design can capitalize on the inherent complexity and instructiveness of natural ECM to build an implant with multi-factorial mechanisms of action that harmonize with healing, serve as a surrogate ECM in the wound, and stimulate the processes of dynamic reciprocity toward renewed homeostasis. Such an implant can guide the patient's cells through a series of cellular recruitment, renewal of lost matrix structures, and reinforcement of tissue strength while undergoing complete turnover and disappearance of the original implant.

The current state of the art for ECM grafts has been described. These materials have shown remarkable success in a wide variety of clinical applications [3, 7]. However, there is still room for improvement. Naturally occurring biologic ECM graft materials can be enhanced or fortified to accelerate some of these biological functions, stimulate cellular phenotype selection, or even create inherent antimicrobial activities that will better withstand infection. Ultimately, the goal of such "next generation" implants must be one of synergizing with natural biology and improving upon the complex interaction of the graft with the patient to allow tissue repair, remodeling, and regenerative processes to proceed unhindered.

Acknowledgements

The authors thank Susan Erb and Samantha Stevenson for their review and edits of the final manuscript.

Conflict of interest

Jason P. Hodde and Michael C. Hiles are employees of Cook Biotech Incorporated and hold multiple patents covering ECM-based biomaterials.

Author details

Jason P. Hodde* and Michael C. Hiles
Cook Biotech Incorporated, West Lafayette, IN, USA

*Address all correspondence to: jason.hodde@cookbiotech.com

References

[1] Gavlin A, Kierans AS, Chen J, Song C, Guniganti P, Mazzariol FS. Imaging and treatment of complications of abdominal and pelvic mesh repair. Radiographics. 2020;40:432-453. DOI: 10.1148/rg.2020190106.

[2] Schneeberger SJ, Kraft CT, Janis JE. No-touch technique of mesh placement in ventral hernia repair: Minimizing postoperative mesh infections. Plast Reconstr Surg. 2020;145:1288-1291. DOI: 10.1097/PRS.0000000000006767.

[3] Cramer MC, Badylak SF. Extracellular matrix-based biomaterials and their influence upon cell behavior. Ann Biomed Eng. 2020;48:2132-2153. DOI: 10.1007/s10439-019-02408-9.

[4] Sandusky GE, Badylak SF, Morff RJ, Johnson WD, Lantz GC. Histologic findings after in vivo placement of small intestinal submucosal vascular grafts and saphenous vein grafts in the carotid artery in dogs. Am J Pathol. 1992;140:317-324.

[5] Liang R, Woo SL, Takakura Y, Moon DK, Jia F, Abramowitch SD. Long-term effects of porcine small intestine submucosa on the healing of medial collateral ligament: A functional tissue engineering study. J Orthop Res. 2006;24:811-819. DOI: 10.1002/jor.20080.

[6] Badylak SF. The extracellular matrix as a biologic scaffold material. Biomaterials. 2007;28:3587-3593. DOI: 10.1016/j.biomaterials.2007.04.043.

[7] Hodde JP. Use of small intestinal submucosa dECM in tissue engineering and regenerative medicine. In: Yamaoka T, Hoshiba T, editors. Decellularized Extracellular Matrix: Characterization, Fabrication and Applications. London, England: The Royal Society of Chemistry; 2020. p. 181-198. DOI: 10.1039/9781788015998.

[8] Ho KL, Witte VMN, Bird ET. 8-ply small intestinal submucosa tension-free sling: Spectrum of postoperative inflammation. J Urol. 2004;171:268-271. DOI: 10.1097/01.ju.0000098680.60020.32.

[9] Soler JA, Gidwani S, Curtis MJ. Early complications from the use of porcine dermal collagen implants (Permacol) as bridging constructs in the repair of massive rotator cuff tears. A report of 4 cases. Acta Orthop Belg. 2007;73:432-436.

[10] Nihsen ES, Johnson CE, Hiles MC. Bioactivity of small intestinal submucosa and oxidized regenerated cellulose/collagen. Adv Skin Wound Care. 2008;21:479-486. DOI: 10.1097/01.ASW.0000323561.14144.19.

[11] Nguyen KP, Zotos V, Hsueh EC. Long-term outcome of biologic graft: A case report. J Med Case Rep. 2014;8:255. DOI: 10.1186/1752-1947-8-255.

[12] Kular JK, Basu S, Sharma RI. The extracellular matrix: Structure, composition, age-related differences, tools for analysis and applications for tissue engineering. J Tissue Eng. 2014;5:2041731414557112. DOI: 10.1177/2041731414557112.

[13] Schultz GS, Davidson JM, Kirsner RS, Bornstein P, Herman IM. Dynamic reciprocity in the wound microenvironment. Wound Rep Regen. 2011;19:134-148. DOI: 10.1111/j.1524-475X.2011.00673.x.

[14] Zhu M, Li W, Dong X, Yuan X, Midgley AC, Chang H, Wang Y, Wang H, Wang K, Ma PX, Wang H, Kong D. In vivo engineered extracellular matrix scaffolds with instructive niches for oriented tissue regeneration. Nat Commun. 2019;10:4620. DOI: 10.1038/s41467-019-12545-3.

[15] Hodde J, Hiles M. Transforming surgery through biomaterial template technology. Br J Hosp Med (Lond). 2016;77:162-166. DOI: 10.12968/hmed. 2016.77.3.162.

[16] Lin PH, Sermersheim M, Li H, Lee PHU, Steinberg SM, Ma J. Zinc in wound healing modulation. Nutrients. 2017;10:16. DOI: 10.3390/nu10010016.

[17] Woo SL, Takakura Y, Liang R, Jia F, Moon DK. Treatment with bioscaffold enhances the fibril morphology and the collagen composition of healing medial collateral ligament in rabbits. Tissue Eng. 2006;12:159-166. DOI: 10.1089/ten.2006.12.159.

[18] Hodde JP. Extracellular matrix as a bioactive material for soft tissue reconstruction. ANZ J Surg. 2006;76:1096-1100. DOI: 10.1111/j.1445-2197.2006.03948.x.

[19] Badylak SF, Park K, Peppas N, McCabe G, Yoder M. Marrow-derived cells populate scaffolds composed of xenogeneic extracellular matrix. Exp Hematol. 2001;29:1310-1318. DOI: 10.1016/s0301-472x(01)00729-9.

[20] Badylak S, Kokini K, Tullius B, Whitson B. Strength over time of a resorbable bioscaffold for body wall repair in a dog model. J Surg Res. 2001;99:282-287. DOI: 10.1006/jsre. 2001.6176.

[21] Franklin ME Jr, Trevino JM, Portillo G, Vela I, Glass JL, Gonzalez JJ. The use of porcine small intestinal submucosa as a prosthetic material for laparoscopic hernia repair in infected and potentially contaminated field: a long term follow-up. Surg Endosc. 2008;22:1941-1946. DOI: 10.1007/s00464-008-0005-y.

[22] Stelly M, Stelly TC. Histology of CorMatrix bioscaffold 5 years after pericardial closure. Ann Thorac Surg. 2013;96:e127-e129. DOI: 10.1016/j. athoracsur.2013.06.114.

[23] Hodde JP, Badylak SF, Shelbourne KD. The effect of range of motion on remodeling of small intestinal submucosa (SIS) when used as an Achilles tendon repair material in the rabbit. Tissue Eng. 1997;3:27-37.

[24] Eming SA, Martin P, Tomic-Canic M. Wound repair and regeneration: mechanisms, signaling, and translation. Sci Transl Med. 2014;6:265sr6. DOI: 10.1126/scitranslmed.3009337.

[25] Korzinskas T, Jung O, Smeets R, Stojanovic S, Najman S, Glenske K, Hahn M, Wenisch S, Schnettler R, Barbeck M. In vivo analysis of the biocompatibility and macrophage response of a non-resorbable PTFE membrane for guided bone regeneration. Int J Mol Sci. 2018;19:2952. DOI: 10.3390/ijms19102952.

[26] Farmer ZL, Domínguez-Robles J, Mancinelli C, Larrañeta E, Lamprou DA. Urogynecological surgical mesh implants: New trends in materials, manufacturing and therapeutic approaches. Int J Pharm. 2020;585:119512. DOI: 10.1016/j. ijpharm.2020.119512.

[27] Harth KC, Blatnik JA, Anderson JM, Jacobs MR, Zeinali F, Rosen MJ. Effect of surgical wound classification on biologic graft performance in complex hernia repair: an experimental study. Surgery. 2013;153:481-492. DOI: 10.1016/j.surg.2012.08.064.

[28] Ayubi FS, Armstrong PJ, Mattia MS, Parker DM. Abdominal wall hernia repair: a comparison of Permacol and Surgisis grafts in a rat hernia model. Hernia. 2008;12:373-378. DOI: 10.1007/s10029-008-0359-z.

[29] Jordan SW, Fligor JE, Janes LE, Dumanian GA. Implant porosity and the foreign body response. Plast Reconstr Surg. 2018;141:103e-112e. DOI: 10.1097/PRS.0000000000003930.

[30] Todros S, Pavan PG, Natali AN. Synthetic surgical meshes used in abdominal wall surgery: Part I-materials and structural conformation. J Biomed Mater Res B Appl Biomater. 2017;105:689-699. DOI: 10.1002/jbm.b.33586.

[31] Fatkhudinov T, Tsedik L, Arutyunyan I, Lokhonina A, Makarov A, Korshunov A, Elchaninov A, Kananykhina E, Vasyukova O, Usman N, Uvarova E, Chuprynin V, Eremina I, Degtyarev D, Sukhikh G. Evaluation of resorbable polydioxanone and polyglycolic acid meshes in a rat model of ventral hernia repair. J Biomed Mater Res B Appl Biomater. 2019;107:652-663. DOI: 10.1002/jbm.b.34158.

[32] Capella-Monsonís H, Tilbury MA, Wall JG, Zeugolis DI. Porcine mesothelium matrix as a biomaterial for wound healing applications. Mater Today Bio. 2020;7:100057. DOI: 10.1016/j.mtbio.2020.100057.

[33] Sun WQ, Xu H, Sandor M, Lombardi J. Process-induced extracellular matrix alterations affect the mechanisms of soft tissue repair and regeneration. J Tissue Eng. 2013;4:2041731413505305. DOI: 10.1177/2041731413505305.

[34] Badylak S, Kokini K, Tullius B, Simmons-Byrd A, Morff R. Morphologic study of small intestinal submucosa as a body wall repair device. J Surg Res. 2002;103:190-202. DOI: 10.1006/jsre.2001.6349.

[35] Clark RA. Fibrin and wound healing. Ann N Y Acad Sci. 2001;936: 355-367. DOI: 10.1111/j.1749-6632.2001.tb03522.x.

[36] Lazaro JL, Izzo V, Meaume S, Davies AH, Lobmann R, Uccioli L. Elevated levels of matrix metalloproteinases and chronic wound healing: an updated review of clinical evidence. J Wound Care. 2016;25:277-287. DOI: 10.12968/jowc.2016.25.5.277.

[37] Hodde JP, Hiles MC, Metzger DW. Characterization of the local wound environment following treatment of chronic leg ulcers with SIS wound matrix. J Tissue Viability. 2020;29:42-47. DOI: 10.1016/j.jtv.2019.12.003.

[38] Przekora A. A concise review on tissue engineered artificial skin grafts for chronic wound treatment: Can we reconstruct functional skin tissue in vitro? Cells. 2020;9:1622. DOI: 10.3390/cells9071622.

[39] Nihsen ES, Zopf DA, Ernst DM, Janis AD, Hiles MC, Johnson C. Absorption of bioactive molecules into OASIS Wound Matrix. Adv Skin Wound Care. 2007;20:541-548. DOI: 10.1097/01.ASW.0000294756.97425.c9.

[40] Witherel CE, Graney PL, Freytes DO, Weingarten MS, Spiller KL. Response of human macrophages to wound matrices in vitro. Wound Repair Regen. 2016;24:514-524. DOI: 10.1111/wrr.12423.

[41] Hodde JP, Ernst DMJ, Hiles MC. An investigation of the long-term bioactivity of endogenous growth factor in Oasis Wound Matrix. J Wound Care. 2005;14:23-25. DOI: 10.12968/jowc.2005.14.1.26721.

[42] McDevitt CA, Wildey GM, Cutrone RM. Transforming growth factor-b1 in a sterilized tissue derived from the pig small intestine submucosa. J Biomed Mater Res. 2003;67A:637-640. DOI: 10.1002/jbm.a.10144.

[43] Record RD, Hillegonds D, Simmons C, Tullius R, Rickey FA, Elmore D, Badylak SF. In vivo degradation of 14C-labeled small intestinal submucosa (SIS) when used

for urinary bladder repair. Biomaterials. 2001;22:2653-2659. DOI: 10.1016/s0142-9612(01)00007-2.

[44] De Silva GS, Krpata DM, Gao Y, Criss CN, Anderson JM, Soltanian HT, Rosen MJ, Novitsky YW. Lack of identifiable biologic behavior in a series of porcine mesh explants. Surgery. 2014;156:183-189. DOI: 10.1016/j.surg.2014.03.011.

Section 2

Hyaluronic Acid and ECM - Interplay and Medicine

Chapter 6

Hyaluronic Acid Fillers: Where We Have Been and Where We Are Going

Alexander Daoud and Robert Weiss

Abstract

Since the approval of the United States' first hyaluronic acid (HA) filler in December 2003, HA fillers have become mainstays of soft tissue augmentation due to their favorable safety profile and minimally invasive treatment nature. The past two decades have not only brought an expansion in the popularity of HA fillers, but also in the number of available HA filler products and indications for cosmetic enhancement. Accordingly, HA filler injection has become one of the most commonly performed cosmetic procedures worldwide. The progression of HA filler products is a study in both biomedical engineering advancements, as well as evolving concepts of beauty and cosmesis. In this chapter, we review the history of these products, including their composition and indications for use. We then explore the prospect of HA fillers for the future of esthetic medicine, as they remain a vital component of nonsurgical soft tissue augmentation.

Keywords: soft tissue augmentation, biomedical engineering, cosmetic dermatology, hyaluronic acid

1. Introduction

Although typically thought of as outgrowths of modern medicine, interventions for esthetic and cosmetic purposes are well documented through multiple civilizations across history. In the non-surgical arena, soft tissue augmentation rapidly accelerated with the first reports of autologous fat transfer in the late 19th century; concomitantly, reports began to emerge from Europe on the use of injectable paraffin for rhytide reduction and soft tissue rejuvenation [1]. Shortly after the spread of paraffin injections across Europe and Asia, reports of embolic complications, as well as late-onset granulomatous reactions (later called 'paraffinomas') led to their eventual removal from the realm of esthetic practice, although non-medical/illicit injection of paraffin-containing substances still remain a sporadic issue in clinical practice today.

It was not until the 1940s that a new injectable agent found widespread use. Silicone, a dimethylsiloxane polymer, first entered clinical practice in Japan as an agent for breast augmentation. Over the following twenty years, silicone found widespread popularity across the United States, with the Dow Corning Corporation developing an injectable form of the material in the mid-1960s. As with paraffin decades prior, reports began to emerge in the 1960s–70s of delayed granulomatous reactions to injected silicone. Termed *siliconomas*, these entities displayed a similar inflammatory pattern to paraffin, although migration of silicone due to the effects of gravity often led to granuloma formation at a site inferior or distal to the original site of injection [2]. As with paraffin, illicit silicone injection remains an issue across the world today, and clinicians should be mindful of this entity in the evaluation of granulomatous reactions following non-medical cosmetic treatments.

In 1981, the cosmetic landscape advanced with the United States Food and Drug Administration's (FDA) approval of Zyderm™, the first injectable filler approved for facial cosmetic enhancement. Derived from bovine collagen, Zyderm™ was comprised of a matrix of type I and type III collagen, which later formulations (Zyplast™) cross-linked with glutaraldehyde in order to slow the degradation of injected material [2]. Unlike the relatively inert nature of both paraffin and silicone, bovine collagen poses a significant risk of hypersensitivity reactions (3–3.5%, per population estimates) due to cross-speciation. Accordingly, skin testing prior to injection was necessary for all patients. Furthermore, although the collagen used for injection was derived from a closed and closely surveilled group of bovines, public concern regarding bovine spongiform encephalopathy ("mad cow disease") led to its fall from favor across the following decades [2].

The modern cosmetic injectables revolution found its genesis in 2003, when the FDA approved Restylane™, its first hyaluronic acid (HA) product. While other non-HA fillers were approved by the FDA for use in the following years – poly-L-lactic acid (Sculptra™), polymethylmethacrylate (Belafill™), and Radiesse™ (calcium hydroxylapatite) among the most commercially successful – HA fillers have risen as some of the most popular agents available for nonsurgical facial cosmesis.

2. Hyaluronic acid fillers: derivatives and bioengineering

Hyaluronic acid is a glycosaminoglycan, a type of acid mucopolysaccharide that demonstrates significant hydrophilicity and serves as a key portion of the extracellular matrix of all organisms. In skin, HA is a key component of "ground substance", the acellular material found in the extracellular environment around collagen bundles of the dermis. In its structural support of collagen and elastin fibers, HA's main role in skin is to lubricate these extracellular structures by attracting water, which in turn produces a volumizing effect in the skin [3].

With time, the amount of HA in the dermis begins to diminish; accordingly, a hallmark of aging skin is a loss of both volume and elasticity. To this effect, HA fillers rose as a safe and logical solution for facial soft tissue enhancement: by reinstating the HA balance of the dermis, HAs draw water into the extracellular environment and lead to significant improvements in rhytide and soft tissue [2, 4]. Furthermore, due to HA's ubiquitous nature as a 'native component' of skin, the immunogenicity of HA filler products remains extremely low.

Engineering of HA fillers is possible through two broad methods: animal-derived and bacterial-derived HAs. In the animal-derived group, HA is extracted from rooster combs and used for cosmetic injection [5]. Among the most successful in this group is Hylaform™, although animal-derived HAs have fallen from popularity due to their relatively shorter duration of effect, as well as their slightly increased immugenicity as compared to bacterial-derived HAs [1, 2]. In the bacterial-derived group, HAs are typically generated by fermentation of nonanimal stabilized hyaluronic acid (NASHA) from streptococcal species (Restylane™, Juvederm™, Captique™, Hydrelle™).

Two unique variables that determine an HA filler's physical properties are gel particle size, as well as the degree of HA crosslinking present. This is best exemplified by the Restalyne™ and Juvederm™ line of HA fillers, where adjustments in these two variables are used to create novel products with different properties and clinical applications. While many other HA filler products are commercially available, further analysis of this line of HA filler agents is chosen due to their ubiquitous presence in cosmetic offices worldwide.

Alterations in gel particle size are a hallmark of the Restylane™ family of products: products featuring particles of lower molecular weight (350 μm for Restylane™ versus 800-900 μm for Restylane™Lyft [formerly known as Perlane]) allow for a higher density of gel particles per unit volume (100,000 particles/milliliter for Restylane™ versus 8,000 particles/milliliter for Restylane™Lyft). This is engineered through a process of sieving – by filtering particles mechanically, products are created with gel particles of a singular size [6]. Accordingly, while small gel particle-based products such as Restylane™, Restylane™-L, Restylane™ Kysse, and Restylane™ Silk have been approved for applications ranging from midface to lip and perioral augmentation, larger gel particle products such as Restylane™ Defyne, Restylane™ Refyne, and Restylane™ Lyft has been typically employed for deep rhytide correction and volumization of the cheeks and midface. Notably, Restylane™ Lyft also carries FDA approval for dorsal hand rejuvenation [1, 6, 7].

The effects of varying degrees of crosslinking are best displayed by the Juvederm™ family of products. Generally speaking, the greater the degree of crosslinking, the greater the degree of water absorption demonstrated by the filler; accordingly, more crosslinked products are typically used when significant volumizing or deep rhytide correction is desired. In contrast to Restalyne, which maintains the same concentration of hyaluronic acid throughout its product line, Juvederm™ products feature varying concentration of hyaluronic acid (24 mg/mL for Juvederm Ultra, Juvederm Ultra Plus, Juvederm Ultra XC, and Juvederm Ultra Plus XC; 20 mg/mL for Juvederm Voluma; 17.5 mg/mL for Juvederm Vollure; 15 mg/mL for Juvederm Volbella). In addition, these products also differ from each other in both their percentage of HA crosslinking, as well as the manner in which they are crosslinked [8, 9].

Juvederm Ultra, Ultra Plus, and their XC variants feature a patented crosslinking technology termed Hylacross. With Hylacross, HAs of the same molecular weight are crosslinked together, creating a relatively uniform product of set viscosity and thickness. This is in contrast to Juvederm Voluma, Vollure, and Volbella, which use another patented technology called Vycross crosslinking. Vycross-generated fillers feature HAs of varying molecular weight, resulting in products that are reported to have a greater number of applications across the face, as well as a longer duration of action due to non-uniform degradation [8, 10].

Together, it is gel particle size and density, as well as degree/type of HA crosslinking, that influence the *G'*, or elastic modulus, of HA fillers. G' is a measure of a filler's response to shear or dynamic forces: the higher the G' of a filler, the greater its resistance to deformation or movement when under the influences of external force (such as dynamic facial movement). Accordingly, fillers with higher G' are often used for their volumizing and lifting effects, and deeper injections are needed [9, 11].

Emblematic of the role of G' in filler applications, Belotero Balance (Merz Esthetics) is a HA filler product that relies on a greater density of non-crosslinked HA in order to produce its intended effect. Generally speaking, the strength and volumizing effect of fillers is directly related to its degree of HA crosslinking – that is, free HAs are non-contributory to the strength, volumizing, or lifting potential of a filler product. Through a novel, patented formulation called Cohesive Polydensified Matrix – a process that generates a varying degree of crosslinked

HA in suspension with free HAs – a product is produced that features a significantly lower G' than other HA products (G' = 30 for Belotero Balance; G' = 545 for Restylane™ Lyft). Accordingly, Belotero ™ Balance is able to be used in both the intradermal and superficial subcuticular planes, with lower risk of the Tyndall effect (a gray-blue discoloration secondary to light scattering from superficially placed filler products) as compared to other HA fillers [9, 11, 12].

Depending on the product used, as well as the location being treated, HA fillers typically augment treated soft tissues for a period of 6–18 months. As compared to bovine collagen, hypersensitivity reactions are exceedingly rare, estimated at 1 in every 5000 injections. A notable benefit of these products, as compared to non-HA fillers, is their reversibility: if dissolution of filler is needed, injectable hyaluronidase may be used.

3. Hyaluronic acid fillers: directions for the future

Given their widespread success and popularity, HA fillers are likely to remain a mainstay in the cosmetic proceduralist's toolbox. Innovations in the field – ranging from methods of biochemical engineering to novel techniques for injection and treatment – all lend great promise to the future of HA fillers in esthetic practice.

In the realm of product development, one such advancement is demonstrated by Teosyal ™, an HA filler approved by the FDA in 2017. Through a novel patented synthesis method, a product with a lower bacterial protein and endotoxin load (as compared to other HA fillers) is generated, with a reported lower risk of potential hypersensitivity reactions as a result [9]. Another product generated through a unique method of synthesis is the Neauvia ™ family of HA fillers. Marketed as a 'fully organic' product, Neauvia ™ HAs are not synthesized from the streptococcal species typical of other HA fillers, but instead by *Bacillus subtilis*, a bacterium widely used in probiotic supplements. These HAs are then crosslinked with polyethylene glycol, creating a biocompatible hydrogel [13]. Lastly, Juvederm ™ has recently introduced Volux, a thicker HA with a concentration of 25 mg/mL, for lower face (jawline and chin) augmentation.

Continuing improvements in injection technique also offer a promising future of HA filler use. Blunt-tipped cannulas are rising in popularity as an alternative to hypodermic needles, as their use is associated with a statistically significant decrease in bruising, as well as lower pain associated with injection [14]. Additionally, cadaveric studies comparing blunt-tipped cannulas to sharp needles demonstrated a higher risk of intra-arterial injection with sharp needles, as well as a greater degree of filler extrusion across multiple anatomic planes (as opposed to one targeted level of injection) with needles as compared to blunt-tipped cannulas [15]. Accordingly, the use of cannulas has greatly advanced the technique, safety profile, and resultant cosmetic effect of HA fillers to areas such as the tear troughs and jawline, where anatomic plane precision is critical.

Exactness in volumetric dosing is highly dependent on injector experience, as elements such as HA filler viscosity, needle gauge, and plane/site of injection all influence the force needed to inject a bolus of filler agent. To elevate this aspect of treatment beyond volume marking on syringes and "injector feel" while introducing a bolus, Restylane ™ has recently introduced Skinboosters, a microdroplet HA treatment that uses a SmartClick® syringe to provide metered doses with each depression of the syringe plunger [16]. This technology has especially shown great promise when injecting into more superficial planes, as improper or non-uniform technique may otherwise result in nodule or 'lump' formation, as well as the Tyndall effect.

4. Conclusions

Hyaluronic acid fillers are essential component of the cosmetic proceduralist's armamentarium. Their ubiquitous presence in clinics across the world, as well as their widespread acceptance by the general public, has made awareness of their use important for all clinicians. An understanding of their unique properties and physical features provides esthetic practitioners with a better comprehension of their treatment indications, potential complications, and treatment pitfalls, as well as an appreciation of developments on the horizon for future HA fillers.

Author details

Alexander Daoud* and Robert Weiss
Maryland Dermatology, Laser, Skin, and Vein Institute, Baltimore, MD, USA

*Address all correspondence to: adaoud@mdlsv.com

References

[1] Kim JE, Sykes JM. Hyaluronic acid fillers: history and overview. Facial Plast Surg. 2011 Dec;27(6):523-528.

[2] Kontis TC, Rivkin A. The history of injectable facial fillers. Facial Plast Surg. 2009;25(2):67-72.

[3] Zerbinati N, Lotti T, Monticelli D, et al. *In Vitro* Evaluation of the Sensitivity of a Hyaluronic Acid PEG Cross-Linked to Bovine Testes Hyaluronidase. Open Access Maced J Med Sci. 2018;6(1):20-24.

[4] Matarasso SL, Carruthers JD, Jewell ML; Restylane Consensus Group. Consensus recommendations for soft-tissue augmentation with nonanimal stabilized hyaluronic acid (Restylane). Plast Reconstr Surg. 2006 Mar;117(3 Suppl):3S-34S; discussion 35S-43S.

[5] Kang DY, Kim WS, Heo IS, Park YH, Lee S. Extraction of hyaluronic acid (HA) from rooster comb and characterization using flow field-flow fractionation (FlFFF) coupled with multiangle light scattering (MALS). J Sep Sci. 2010;33(22):3530-3536.

[6] Khosravani N, Weber L, Patel R, Patel A. The 5-Step Filler Hand Rejuvenation: Filling with Hyaluronic Acid. Plast Reconstr Surg Glob Open. 2019;7(1):e2073.

[7] Dover JS, Rubin MG, Bhatia AC. Review of the efficacy, durability, and safety data of two nonanimal stabilized hyaluronic acid fillers from a prospective, randomized, comparative, multicenter study. Dermatol Surg. 2009 Feb;35 Suppl 1:322-330; discussion 330-1.

[8] Bogdan Allemann I, Baumann L. Hyaluronic acid gel (Juvéderm) preparations in the treatment of facial wrinkles and folds. Clin Interv Aging. 2008;3(4):629-634.

[9] Rohrich RJ, Bartlett EL, Dayan E. Practical Approach and Safety of Hyaluronic Acid Fillers. Plast Reconstr Surg Glob Open. 2019;7(6):e2172.

[10] Micheels P, Besse S, Sarazin D. Two Crosslinking Technologies for Superficial Reticular Dermis Injection: A Comparative Ultrasound and Histologic Study. J Clin Aesthet Dermatol. 2017;10(1):29-36.

[11] La Gatta A, Salzillo R, Catalano C, et al. Hyaluronan-based hydrogels as dermal fillers: The biophysical properties that translate into a "volumetric" effect. *PLoS One*. 2019;14(6):e0218287. Published 2019 Jun 11.

[12] Sundaram H, Fagien S. Cohesive Polydensified Matrix Hyaluronic Acid for Fine Lines. Plast Reconstr Surg. 2015 Nov;136(5 Suppl):149S-163S.

[13] Zerbinati N, Mocchi R, Galadari H, et al. *In Vitro* Evaluation of the Biological Availability of Hyaluronic Acid Polyethylene Glycols-Cross-Linked Hydrogels to Bovine Testes Hyaluronidase. *Biomed Res Int*. 2019;2019:3196723. Published 2019 Jun 12.

[14] Fulton J, Caperton C, Weinkle S, Dewandre L. Filler injections with the blunt-tip microcannula. J Drugs Dermatol. 2012;11(9):1098-1103.

[15] van Loghem JAJ, Humzah D, Kerscher M. Cannula Versus Sharp Needle for Placement of Soft Tissue Fillers: An Observational Cadaver Study. Aesthet Surg J. 2017 Dec 13;38(1):73-88.

[16] Lee BM, Han DG, Choi WS. Rejuvenating Effects of Facial Hydrofilling using Restylane Vital. Arch Plast Surg. 2015;42(3):282-287.

Chapter 7

Hyaluronic Acid Derivatives for Targeted Cancer Therapy

Nilkamal Pramanik and Sameer Kumar Jagirdar

Abstract

Targeted therapeutics are considered next generation cancer therapy because they overcome many limitations of traditional chemotherapy. Cancerous cells may be targeted by various hyaluronic acid modified nanovehicles that kill these cells. Particularly, hyaluronic acid and its derivatives bind with high affinity to cell surface protein, CD44 enriched tumor cells. Moreover, these molecules have the added advantage of being biocompatible and biodegradable, and may be conjugated with a variety of drugs and drug carriers for developing various formulations as anti-cancer therapies such as nanogels, self-assembled and metallic nanoparticulates. In this chapter, we have covered various aspects of hyaluronic acid-modified delivery systems including strategies for synthesis, characterization, and biocompatibility. Next, the use of hyaluronic acid-modified systems as anti-cancer therapies is discussed. Finally, the delivery of small molecules, and other pharmaceutical agents are also elaborated in this chapter.

Keywords: Hyaluronic Acid, Nano-particulates, Immunogenicity, Biodegradation, Tumor Targeted delivery

1. Introduction

Nanoparticles have gained increased attention in the context of cancer therapy; however, the major challenge of targeting particles specifically to cancerous cells remains. Targeted delivery systems comprising cell-targeting ligands such as antibodies, peptides, folic acid, and various biomolecules have been developed to ensure tumor-specific delivery. One such targeting ligand is hyaluronic acid (HA).

HA, a glycosaminoglycan (GAG), is a natural polysaccharide present in the extracellular matrix of various soft connective tissues such as the vitreous humor, dermis of the skin, hyaline cartilage, and synovial fluid of the body. It is water-soluble, viscoelastic, biodegradable, biocompatible, and non-immunogenic [1–4]. HA is a polyanionic mucopolysaccharide consisting of β-1,3 and β-1,4 glycosidic bonds between repeating units of D-glucuronic acid and N-acetyl-D-glucosamine [5].

As an intrinsic part of the ECM, HA participates in different biological functions of the cell, including signal transduction, vascularization, cell migration, and tissue remodeling as schematically represented in **Figure 1**. Additionally, the presence of modifiable hydrophilic functional groups—hydroxyl, carboxyl, and N-acetyl increases its potential as an adaptable system for the delivery of proteins, nucleic acids, and anti-cancer agents by grafting or modification with different nanoparticles. Based on cellular interaction studies, HA has emerged as a tumor-targeting agent in cancer therapy. It exhibits a

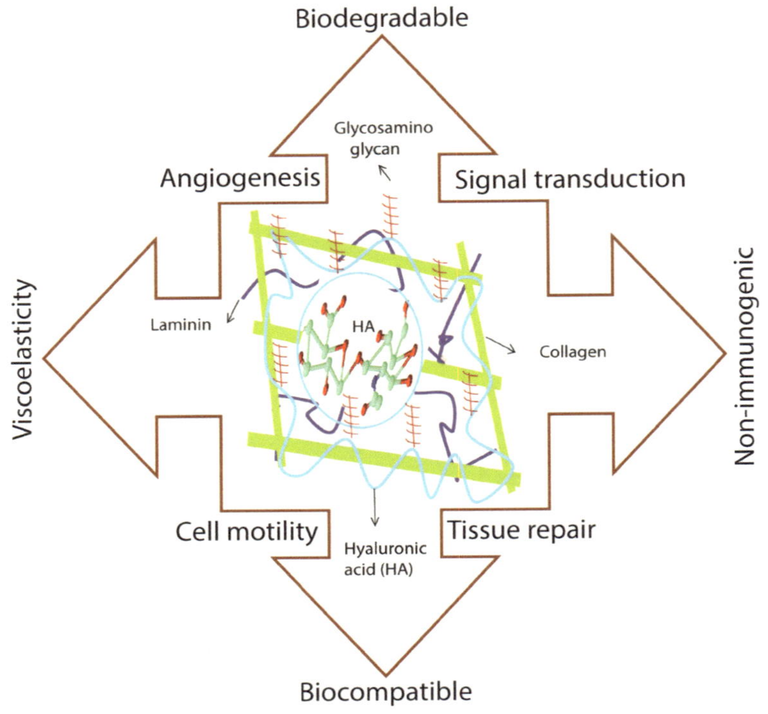

Figure 1.
Schematic representation of physico-chemical properties and functions of hyaluronic acid (HA) in the native tissue environment.

high binding affinity towards the CD44 cluster [6], which is over-expressed in numerous malignant cancer cells [7–9].

Considering its hierarchical structure and potential as a targeting agent, HA can be modified using small biomolecules [10], nanotubes [11], or different types of metal and non-metal nanoparticles [12] in various formulations. HA modified particles are effectively ingested by cells, and HA is enzymatically degraded intracellularly [13], resulting in the delivery of only the particles and their cargo inside the cell.

This chapter covers various aspects of HA-modified delivery systems, including strategies for synthesis, characterization, and biocompatibility. Furthermore, the characteristics of different delivery systems such as nanogel, micelle, liposome, and metallic or non-metallic nano-particulates are discussed. The mechanism underlying the delivery of small molecules, nucleic acids, and other pharmaceutical agents (*in vitro*, *in vivo*, or clinical applicability) are also presented in this chapter.

2. Hyaluronic acid–a primer

HA was discovered in 1934 when scientists Karl Meyer and John Palmer isolated a new kind of polysaccharide from bovine vitreous humor, which was later termed as hyaluronan [14]. Hyaluronic acid is a polysaccharide composed of repeating units of β-1,3-N-acetyl-D-glucosamine and β-1,4-D-glucuronic acid linked by β-1,3 and β-1,4 glycosidic bonds. It is a component of the extracellular matrix and mediates various cellular functions. HA is a ligand for the cell surface receptor CD44. CD44 is a transmembrane protein that is typically expressed by a number of cells, but is over-expressed in different types of metastatic tumor cells, including those of brain, breast, prostate, colon, bladder, and head and neck cancers [7–9, 15–17]. It has a significant role in cell proliferation, migration, metastasis, and cell–cell and cell-matrix signal transduction [18].

2.1 Synthesis

Hyaluronic acid is usually obtained from different biological sources. The microbial synthesis pathway is preferred as it is cost-effective and an environmentally benign process. The gram-positive bacterium, *Streptococcus zooepidemicus* is used for large-scale production of HA via a fermentative pathway as shown in **Figure 2(A)** [19]. The biosynthetic pathway of HA is as follows: initially, glucose-6-phosphate is converted to uridine diphosphate glucose (UDP-glucose) in the presence of α-phosphoglucomutase and UDP-glucose dehydrogenase followed by UDP-glucose dehydrogenase assisted oxidation into UDP-glucuronic acid. Next, an amide group is transferred from glutamine-fructose-6-phosphate to fructose-6-phosphate, followed by rearrangement of the phosphate leading to the formation of glucosamine-1-phosphate. Subsequently, acetylation and conjugation of UTP to glucosamine-1-phosphate generates the second precursor of HA. In the final step, hyaluronan synthase polymerizes the two precursors to produce HA, which is presented as an extracellular capsule as confirmed by electron microscopy (see **Figure 2(B)**) [20]. However, the pathogenicity of the bacterium limits its use for mass production of HA and therefore, several recombinant strains such as *Agrobacterium* sp. ATCC 31749 and recombinant *Escherichia coli* have been adapted as an alternative source of HA production [21, 22].

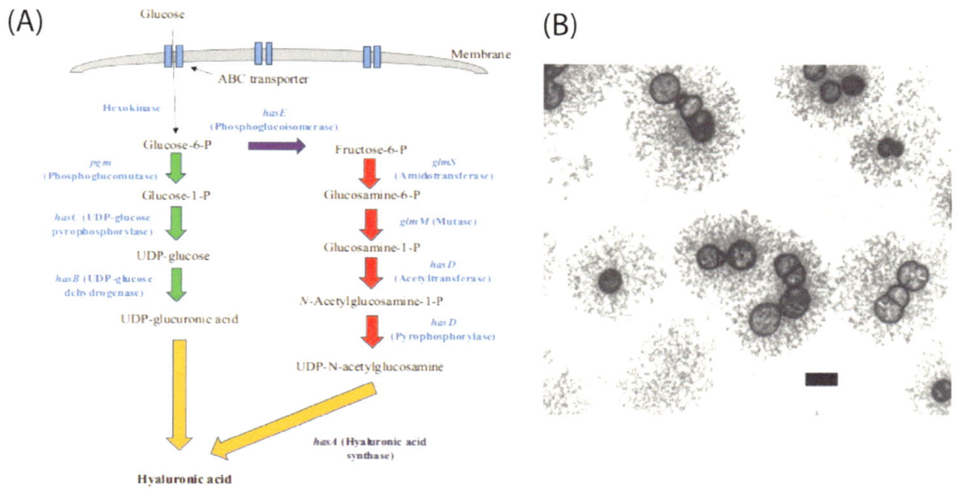

Figure 2.
(A) Hyaluronic acid biosynthetic pathway in S. zooepidemicus*. Glucose is first converted to glucose-6-phosphate by hexokinase which then enters one of two distinct pathways to form UDP-glucuronic acid (*pgm, hasC *and* hasB*) or UDP-*N*-acetylglucosamine (*hasE, glmS, glmM *and* hasD*). These precursors are subsequently bound together via the action of hyaluronic acid synthase or HAVE (encoded by* hasA *in* S. zooepidemicus*) to form hyaluronic acid. [Ref. [19], reproduced with permission from publishing authority]. (B) Electron micrograph section of* Streptococcus equi *subsp.* Zooepidemicus (S. zooepidemicus) *cells obtained from the late exponential phase of an aerated bioreactor culture. Thin sections were stained with uranyl acetate and lead citrate and examined with a Jeol JEM-1010 transmission electron microscope at an accelerating voltage of 80 kV.* Bar 1 *μm. [Ref. [20], reproduced with permission from publishing authority].*

2.2 Properties and uses

As mentioned previously, HA is a polymer. The physiochemical properties such as rheology and viscoelasticity of the polymer depend on the length of HA. Previous studies have demonstrated that higher molecular weight HA has better wound healing properties and is more effective for orthopedic treatment whereas low molecular weight HA has a prominent role in angiogenesis and is an effective immuno-stimulant [23]. A variety of HA oligosaccharides have been synthesized chemically, ranging from disaccharides to hexasaccharides to improve biomedical availability and use. Lu et al. synthesized HA decasaccharides using a chemoselective glycosylation pathway [24]. Commercially available D-glucose and D-glucosamine hydrochloride are chemically connected in the presence of an activator to generate various HA-derivatives [25].

3. Applications in cancer therapy

On the basis of their physiochemical properties, different HA-based nanoformulations have been investigated for their therapeutic application in tumortherapy. HA conjugated drugs, polymers, and lipids may self-assemble in aqueous solvents, and this property has been used to synthesize a number of self-assembled nano-particulates that are either made of the drug or contain the drug.

3.1 Drug-HA conjugates

Direct conjugation of drug molecules with HA results in the development of systems that are not only capable of targeting but improve solubility as well as blood circulation times of the drug itself. Some of the methods to synthesize drug-HA nanoparticles are summarized in **Figure 3**, and below, we discuss a few examples of the use of HA in developing new anti-cancer therapeutics.

The first example, presented in **Figure 3(A)**, involves the conjugation of HA with succinic anhydride derived paclitaxel (2-succPTX, a tubulin inhibitor) using glutathione (GSH) sensitive cystamine (or non-sensitive adipic dihydrazide) as a cross-linker. The resultant self-assembled nanoparticles formed were analyzed using ^{1}H nuclear magnetic resonance (NMR), Fourier-transform infrared spectroscopy (FTIR), and UV–visible spectroscopy. Transmission electron microscopy (TEM) and atomic force microscopy (AFM) analysis demonstrated the presence of spherical shaped nanoparticles of diameter 150 nm. The nanoparticles accurately targeted cancer cells with significant anti-tumor efficacy both *in vitro* and *in vivo* as compared to free PTX [26]. Along similar lines, another HA-PTX nanovehicle showed excellent results in reducing tumor size with the increase of survival rate in an *in vivo* mouse xenograft model bearing ovarian cancer cells [27]. A recent development in this area was to improve the loading of PTX in these self-assembled nanoparticles through the use of dimethylsulfoxide (DMSO) and polyethylene glycol (PEG) in the organic phase. This nano-system was suggested to have a 10–20% increase in PTX loading, and as a result, had significant anti-tumor activity against the RT-4 and RT-112/84 bladder carcinoma cell-lines [28].

The second example is the synthesis of a HA-doxorubicin (DOX) based self-assembled pro-drug that formed spherical core-shell nanostructures (**Figure 3(B)**) of 180–200 nm diameter. It displayed good biocompatibility and pH-responsive controlled Dox release in a cervical cancer model and exhibited excellent tumor inhibitory effects [29]. An ion-pairing based Dox-HA nano-assembled structure

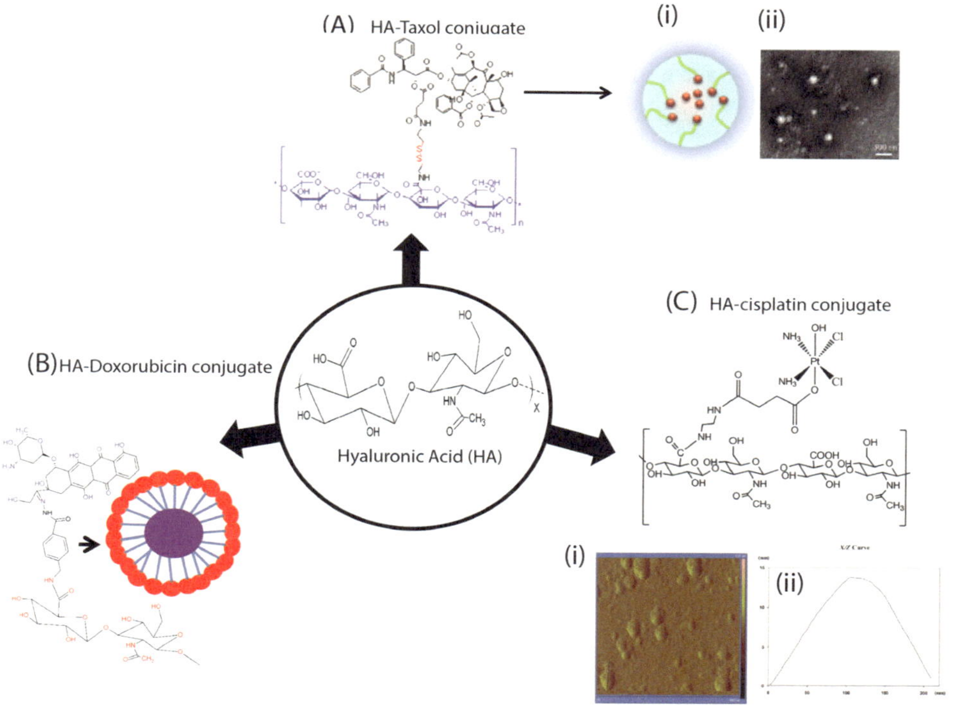

Figure 3.
(A) Synthesis of 2′-succinyalted paclitaxel (PTX) and cystamine (ss) modified hyaluronic acid (HA)-g-paclitaxel (PTX) [HA–ss–PTX] through two consecutive pathways. Initially, PTX was functionalized with succinic anhydrate to obtain an active free carboxylic acid group which was further grafted with HA in the presence of cystamine as a cross-linker using EDAC. HCl and NHS reaction chemistry mechanism. [A(i&ii)] illustrate the schematic and TEM images of HA-PTX based self-assembled structure. [Ref. [26], adapted with permission from publishing authority]. (B) Schematic presentation of the synthesis of amide methyl 4 (aminomethyl) benzoate crosslinked with self-assembled HA-doxorubicin graft. (C) Synthesis of 2, 7-succinyalted cisplatin [2, 7-Succ-Pt(IV)] followed by ethylene diamine (EDA) mediated conjugation with HA using EDAC. HCl and NHS reaction chemistry to form HA-EDA-Pt(IV) pro-drug. [C(i&ii)] AFM images of HA–EDA–Pt (IV) nanoconjugate at an optimal dilution ratio, indicating the formation of microspheres with an average diameter of 200 nm. [Ref. [31], adapted with permission from publishing authority].

with a liposomal delivery system displayed sustained intracellular release of Dox in CD44+ cancer cells with enhanced therapeutic efficacy in a mouse model [30].

The third and final example is the conjugation of HA with cisplatin. Cisplatin is an anti-cancer drug that causes adverse side-effects. Increasing cancer cell-specific intracellular delivery of cisplatin may reduce side effects, which might be possible by conjugating HA onto the drug. Ling et al. have demonstrated the synthesis of HA conjugated cisplatin using N-(3-dimethylaminopropyl)-N′-ethylcarbodiimide hydrochloride. The synthesized pro-drug was verified using ^{1}HNMR, ^{13}C NMR, FT-NIR, AFM, and DSC analysis. The prepared pro-drug was spherical (**Figure 3(C)**) and showed CD44 mediated endocytosis with negligible stimulation to blood vessels. Systemic toxicity studies indicated that the drug was safe and actively delivered cisplatin to kill tumor cells with reduced adverse effect on healthy cells [31].

3.2 Drug loaded HA based nanoparticles

HA may also be conjugated with lipids and other small molecules, which also self-assemble to form particulates that may be used to encapsulate drugs. One example of such a system is deoxycholic acid conjugated with HA, which results in the formation of micelles that may, in turn, be loaded with drug molecules (**Figure 4(A)**). In the specific study presented in this figure, Huo and colleagues showed that the average hydrodynamic size and colloidal stability in terms of 'zeta potential' of the resultant micelle was 120 nm and 36 mV, respectively. These particles were capable of releasing taxol (the drug here) into the cytoplasm of cancer cells, causing tumor apoptosis, with a minimum adverse effect on healthy cells [32]. In another study, cholesterol coupled HA was employed as the amphiphilic molecule that self-assembles into nanoparticles that encapsulate both the anti-cancer drug Dox and magnetic nanoparticles (**Figure 4(B)**). This multifunctional delivery system exhibited high cytotoxicity and cellular uptake against several cancer cell lines such as HeLa, HepG2, and MCF7 [33].

Utilizing HA-molecule conjugated self-assembled nanovehicles for drug loading enables the development of combinatorial therapeutic approaches. As an example of such an approach, the photo-sensitizer, Ce6, has been coupled with HA to form a self-aggregated system that is capable of delivering therapeutic drugs. Such a system was used in a human colon xenograft and displayed a significant anti-tumor effect

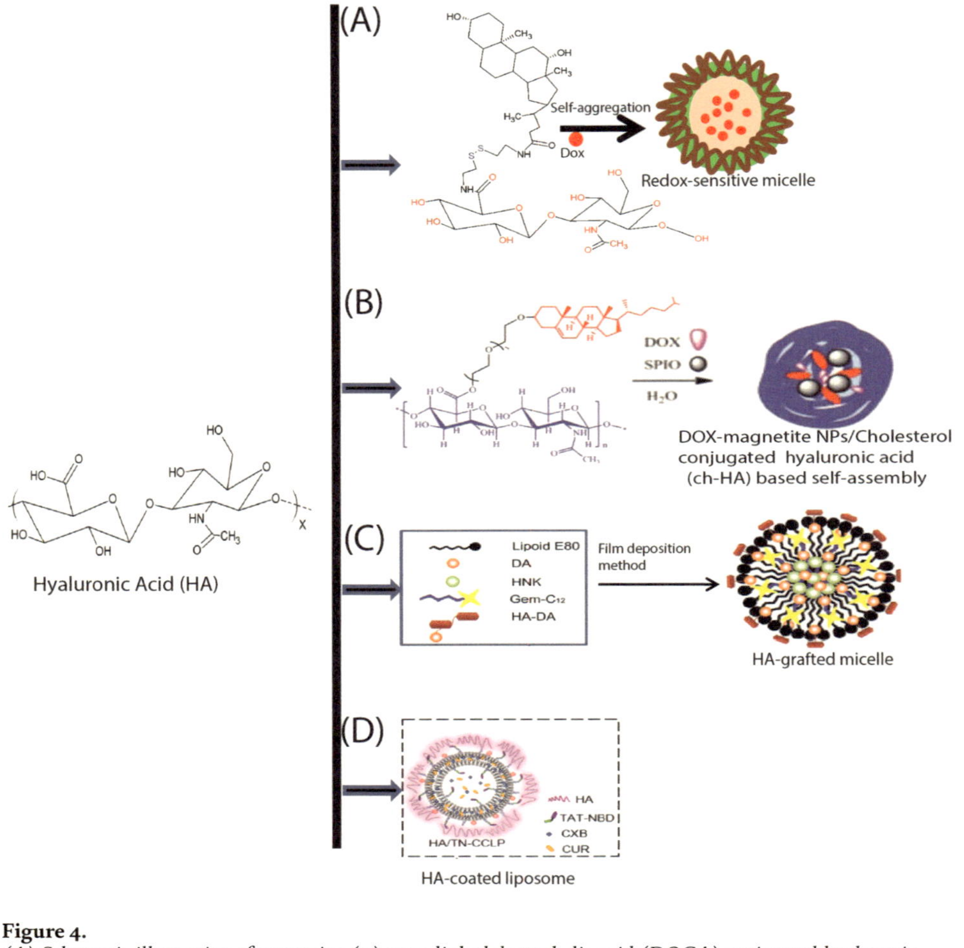

Figure 4.
(A) Schematic illustration of cystamine (ss) cross-linked deoxycholic acid (DOCA) conjugated hyaluronic acid (HA) [HA-ss-DOCA conjugate] nanoparticles. (B) Schematic illustration of the synthesis of cholesterol-conjugated HA (ch-HA) and the formation of DOX/SPION loaded ch-HA micelles. [Ref. [33], adapted with the permission from publishing authority]. (C) Schematic illustration of the preparation of HA-grafted micelles (HA-M) [Ref. [36], reproduced with permission from publishing authority]. (D) Preparation of HA/TN-CCLP and other liposomes. [Ref. [37], adapted with permission from publishing authority.

in a mouse model [34, 35]. In the case of brain tumor therapy, the delivery of chemotherapeutic agents is hindered by the sophisticated blood–brain barrier (BBB) [36], and such combinatorial therapeutics could be beneficial for hard-to-treat cancers such as glioblastoma multiforme (GBM). Combinatorial chemotherapy strategies based on lauroyl-gemcitabine, honokiol (HNK), and HA grafted micelle formulations (**Figure 4(C)**) were shown to significantly suppress GBM in an *in vivo* xenograft model [36]. Similarly, Liposomes composed of TAT-NBD (TN, a 22 amino acid cell-penetrating peptide) modified HA, encapsulating celecoxib (CXB) and curcumin (CUR) (HA/TN-CCLP) (**Figure 4(D)**) were reported to block nuclear factor-κB (NF-κB) and signal transducer and activator of transcription 3(STAT3) signaling pathways, potentially inhibiting tumor growth and metastasis by improving infiltration of inflammatory cells [37].

3.3 Gel formulations

One of the major advantages of using HA is its ability to be used in diverse forms. As HA is an extracellular matrix protein, it may also be used to form hydrogels by itself or in combination with other polysaccharides. Injectable polymeric hydrogels have made a significant contribution to active targeted delivery of chemotherapeutic agents as the 3-dimensional porous environment allows for pH or thermo-sensitive controlled intracellular release of cargo. A HA modified chitosan grafted poly-N-isopropylacrylamide hydrogel was reported to have the loading capacity of Dox/folic acid-g-graphene sheets with high killing efficacy against MCF7 breast cancer cells. An *in vivo* study also demonstrated delivery of anti-tumor agents using the same system [38].

Interferon α-2a (IFN α-2a) loaded HA–tyramine hydrogels have also been shown to have anti-tumor effects, while native IFN α-2a injection did not show any anti-cancer effects. This was due to the controlled release of IFN α-2a from the hydrogel network [39]. Such gels may also be developed as nano-formulations, as demonstrated by Jaya Kumar and colleagues who showed that a redox-sensitive Dox loaded chitin-cystamine-HA nanogel may be used to specifically kill CD44+ HT-29 cells [40].

3.4 Graphene oxide (GO) based formulations

In the past couple of decades, carbonaceous compounds have gained significant attention in cancer therapy due to their large surface area and bio-sensing, bio-imaging, cellular probing, and drug carrier abilities. Owing to its 2D structure, biocompatibility, and water-dispersion features, graphene oxide (GO) and its HA conjugate have been investigated as anti-cancer drug delivery platforms. Three specific examples are discussed here.

First is the development of HA and Arg-Gly-Asp (RGD) peptide coated graphene oxide as a nano-carrier for Dox. Raman spectroscopic analysis revealed two strong peaks at 1350 cm^{-1} and 1550 cm^{-1} due to the presence of D and G bands in exfoliated graphene oxide, as shown in **Figure 5(A)**, which was further confirmed by TEM. The nano-carrier was composed of a single transparent layer with a gauze-like lid layer. It exhibited high Dox loading capacity and excellent cytotoxicity when tested on an ovarian cancer cell line, SKOV-3. It was also found to be biocompatible when tested on a healthy human cell line, HOSEpiC [41].

The second is the development of a redox-sensitive near-infrared (NIR) controlled system consisting of both HA and GO. Yin et al. demonstrated (**Figure 5(B)**) that a Dox/HA-cystamine-GO nano-carrier displays selective targeting and glutathione responsive release of Dox in the cytosol without any collateral damage to healthy

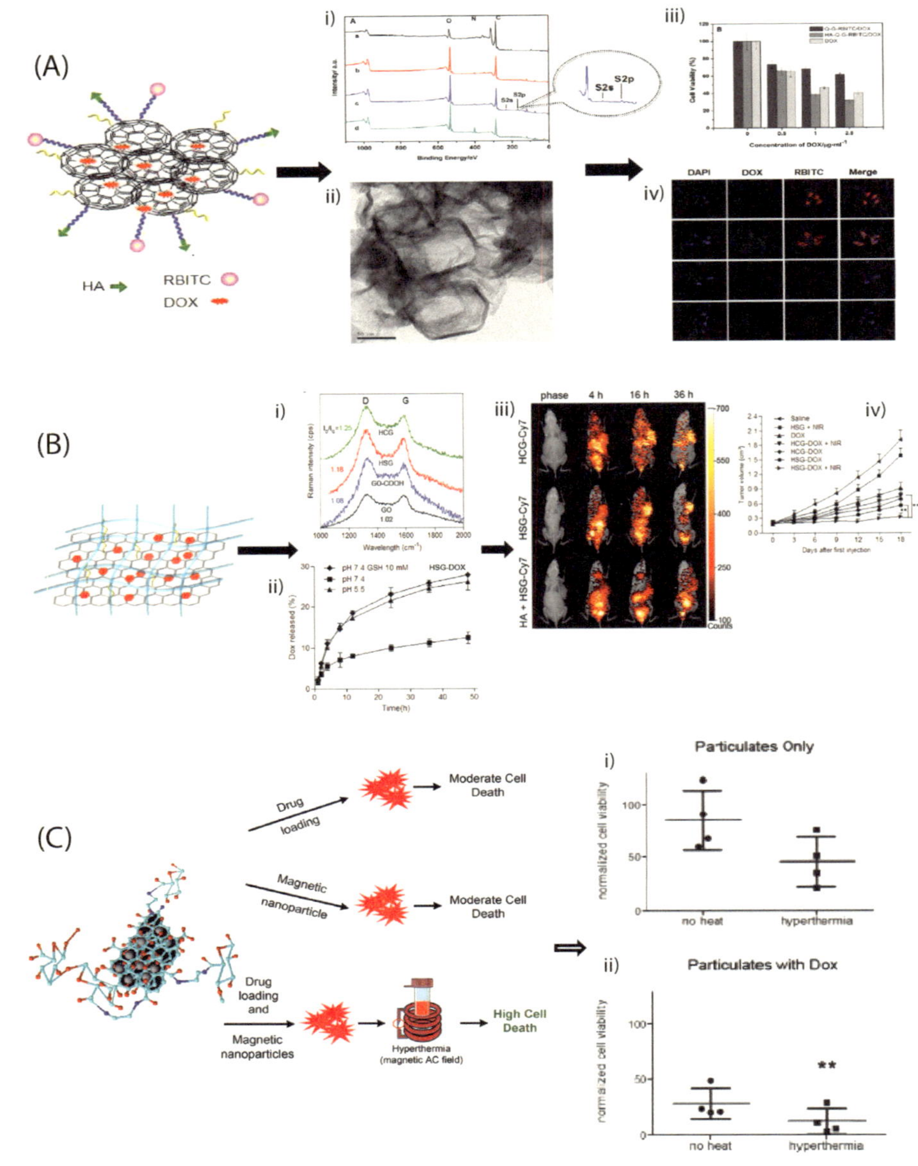

Figure 5.

*(A) Schematic illustration of the preparation and characterization of RBITC labeled Q-graphene (HA-Q-G-RBITC)/DOX nanoparticles and HA-mediated endocytosis. (i) XPS analysis of Q-graphene (a), Q-graphene-COOH (b), Sulf-Q-graphene (c), and PEG Q-graphene (d). (ii) TEM image of Q-graphene. (iii) cytotoxic effects of Q-G-RBITC/DOX, HA-Q-G-RBITC/DOX, and DOX against A549 cells with increasing DOX concentration. The error bar represents standard deviation (n = 5). (iv) CLSM images of A549 cells incubated with (A) HA-Q-G-RBITC, (B) HAQ-G-RBITC/DOX, or (C) Q-G-RBITC/DOX for 5 h. (D) MRC-5 cells incubated with HA-Q-G-RBITC/DOX for 5 h. [Ref. [41], reproduced with permission from publishing authority]. (B) Synthesis of HSG-DOX nanosheets. (i) Raman absorption spectra of HCG, HSG, GO-COOH, and GO. (ii) In vitro DOX release from HSG-DOX after incubation with glutathione at 37°C. (iii) In vivo fluorescent imaging of MDA-MB-231 tumor-bearing nude mice at 4, 16, and 36 h after intravenous injection of Cy7-labeledHSG-DOX and HCG-DOX nanosheets with or without pre-injection of free HA at HSG/HCG dose of 5 mg/kg. (iv) tumor growth curves after intravenous injection of different formulations at a DOX dose of 5 mg/kg. **P < 0.01. [Ref. [12], reproduced with permission from publishing authority]. (C) Synthesis of dox loaded magnetite nanoparticles/GO-HA. Cytotoxicity induced by combination of drug treatment and hyperthermia. GO-HA-iron oxide particulates (i) without or (ii) with dox were cultured with MDA-MB-231 cells and exposed to magnetic fields, and cell viability was measured. Cell viability measurements are normalized to control cultures of cells in the absence of particulates, drugs, and hyperthermia. Paired Student's t-test was performed to compare hyperthermia treatment to their respective no-heat controls. **p < 0.01. [Ref. [43], adapted with permission from publishing authority].*

cells. As compared to free Dox, the NIR irradiated nanosystem exhibited enhanced cytotoxicity in a xenograft tumor model [12]. A similar cytotoxic effect was observed when HiLyte 647 loaded nano GO-HA was used to treat melanoma. Photo-thermal treatment resulted in the complete ablation of tumor tissue without any further tumorigenesis [42].

The third example is our own work on a formulation containing magnetic nanoparticle decorated GO-HA, which was evaluated for magnetothermal and CD44 (+) positive breast cancer targeted cancer therapy. A shown in **Figure 5(C)**, the nanoplatform can be loaded with various types of chemotherapeutic agents such as Dox and Ptx. Furthermore, the study revealed that the nanoplatform had significant anti-tumor activity under magnetic hyperthermia in the MDA MB231 cell line. These nanovehicles provide a versatile platform for next-generation cancer therapy [43].

3.5 Other carriers

Owing to their high stability, biocompatibility, and tunable porous architecture, mesoporous silica nanoparticles (MSNPs) have been used as multifunctional tumor-targeting nano-carriers. Dox loaded HA modified MSNPs have been developed and tested against HCT-116 cells, as shown in **Figure 6(A)**. As part of the morphological analysis, TEM revealed spherical nanoparticles organized as a hexagonally packed mesoporous structure with a mean particle size of 70–100 nm. Surface modification of the MSNPs was confirmed by ^{13}C NMR analysis. Strong absorption peaks at 43, 22, and 10 ppm and broad peaks at 70–180 ppm confirmed the presence of the methylene carbon in NH_2-MSNPs and the anomeric carbon in HA, respectively. Compared to free Dox and Dox loaded MSNPs, Dox-HA-MSNPs exhibited a significantly greater anti-proliferative effect because of better CD44 mediated uptake of HA modified nanoparticles at physiological pH [44].

A similar approach was used to fabricate mesoporous silica nanoparticles, post-functionalized with PEG-PDS-NH2 [poly(poly(ethyleneglycol) methacrylate-co-pyridyldithioethyl methacrylate-co-2 aminoethylmethacrylate], followed by HA decoration for selective targeting. As shown in **Figure 6(B)**, Dox loaded nanoparticles demonstrated clathrin and macropinocytosis-mediated cellular uptake with the killing of CD44 positive HeLa cells [45].

Gold nanoparticles (AuNPs) in cancer therapy have dual functionality due to the presence of a bioactive surface, contrast ability, and photodynamic features. Kang et al. developed HA conjugated pheophorbide-A coated AuNPs (**Figure 6(C)**) that had excellent colloidal stability and photoactivity in the intracellular environment [46]. Electron microscopy analysis of the hybrid nanomaterial revealed spherical nanoparticles with a mean diameter of 70–80 nm, whereas native gold nanoparticles had a mean diameter of 10–15 nm. The increase in size is attributed to the surface coating of the AuNPs. Active targeting was observed 48 h post-injection of PheoA-HA/AuNPs as indicated by bright fluorescence intensity at the tumor site rather than elsewhere, suggesting CD44 receptor-mediated accumulation of nanoparticles with minimum adverse effects on healthy tissues. *In vivo* study of administration of PheoA-HA/AuNPs revealed their excellent anti-tumor efficacy with 3-fold to 5-fold decrease in tumor size compared to free PheoA, saline, and AuNPs. In another study, Wang et al. synthesized {(Au0)100G5.NH2-FI-DOTA (Mn)-HA, where DOTA-1, 4, 7, 10-tetraazacyclododecane-1, 4, 7, 10-tetraacetic acid} NPs which facilitated selective internalization of dendrimers in tumor cells whose imaging capability is presented in **Figure 6(D)** [47].

Super-paramagnetic iron oxide nanoparticles have gained tremendous attention in targeted cancer therapy due to their diverse properties, including

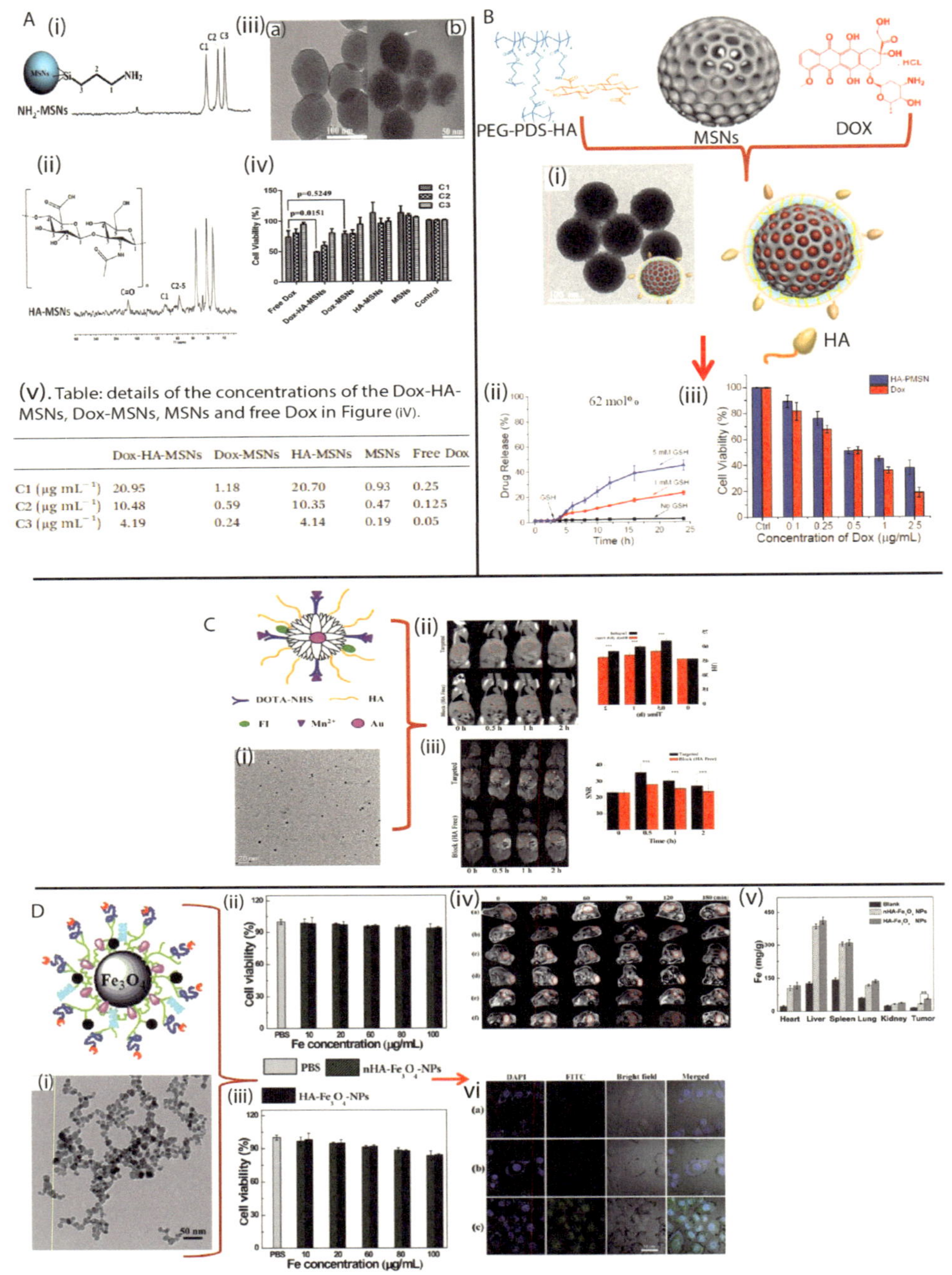

	Dox-HA-MSNs	Dox-MSNs	HA-MSNs	MSNs	Free Dox
C1 (μg mL^{-1})	20.95	1.18	20.70	0.93	0.25
C2 (μg mL^{-1})	10.48	0.59	10.35	0.47	0.125
C3 (μg mL^{-1})	4.19	0.24	4.14	0.19	0.05

Figure 6.

(A) Synthesis of different hyaluronic acid (HA) modified metal nano formulations for targeted cancer therapy. [A(i&ii)]. ^{13}CNMR spectra of NH2-MSNs and HA-MSNs. [iii(a&b)]. TEM images of MSNs and (iv) HA-MSNs. (D) Cytotoxicity of free dox, dox-HA-MSNs, dox-MSNs, HA-MSNs and MSNs againstHCT-116 cells at different concentrationsC1, C2andC3 (for details, see table v). [Ref. [44], adapted with permission from publishing authority]. (B) Schematic illustration of efficient mesoporous nanoparticle-mediated DDSs using noncovalent polymer gatekeepers and HA conjugation for targeting capability. (i) TEM images of HA-PMSNs, (ii) cumulative dox release profiles of 62 Mol% crosslinked PMSNs with (1 and 5 mM) and without GSH. (iii) cell viability analysis of HA-PMSNsCD44-positive in HeLa cells. [Ref. [45], adapted with permission from publishing authority]. (C) Schematic representation of the synthesis of {(Au0)100G5. NH2-FI-DOTA (Mn)-HA} NPs. (i) TEM image of the {(Au0)100G5.NH2-FI-DOTA (Mn)-HA} NPs. The scale bar in each panel measures 20 nm. (ii) In vivo CT images of orthotopic liver tumors at different times after a 0.3-mL intravenous injection of a {(Au0)100G5.NH2-FI-DOTA (Mn)-HA} NP solution (0.3 mL

in PBS, [Au] = 120 mM). (iii) In vivo MR images of orthotopic liver tumors at different times after an intravenous injection of 0.3 mL of a {(Auo)100G5.NH2-FI-DOTA (Mn)-HA} NP (300 μg Mn) solution in PBS. [Ref. [47], adapted the permission from publishing authority]. (D) Synthesis of HA-modified magnetite nanoparticles (HA-Fe3O4 NPs) (i) TEM micrographs of HA-Fe3O4 NPs. The viability of MIAPaCa-2 cells after treated with PBS, nHA-Fe3O4 NP and HA-Fe3O4 NPs at the different Fe concentrations for 24 h (ii) or 48 h (iii) at 37 oC by the CCK-8 assay (iv) In vivo transverse T2 MR images of tumors after intravenous injection of the nHA-Fe3O4 NP ((a) 7 days; (c) 14 days; (e) 21 days) and HAFe3O4 NPs ((b) 7 days; (d) 14 days; (f) 21 days) ([Fe] 1 mg/mL, in 200 mL saline) at different time points post i.v.-injection. (v) In vivo biodistribution of hearts, livers, spleens, lungs, kidneys, and tumors 24 h post intravenous injection of the nHA-Fe3O4 NP and HA-Fe3O4 NPs (600 mg Fe, in 0.3 mL PBS). (vi) The ability of MIAPaCa-2 cells to uptake PBS (a), nHA-Fe3O4 NP (b) and HA-Fe3O4 NPs (c) ([Fe] 50 mg/mL) 4 hours after treatment, MIAPaCa-2 cells treated with PBS were used as control, scale bar ¼ 10 mm. [Ref. [48], adapted with permission from publishing authority].

biocompatibility, enabling their use as a contrast agent for MRI and in magneto-thermal therapy. Moreover, their tendency for aggregation in an aqueous medium is avoided by coating with active biopolymers and cancer-targeting agents. For example, HA modified and fluorescein isothiocyanate decorated iron nanoparticles may be utilized as an efficient probe for targeted MRI assisted cancer therapy, as shown in **Figure 6(E)** [48].

Another use of these particulates could be targeted brain tumor therapy, and a specific example is the development of HA-polyethylene glycol stabilized magnetite nanoparticle modified nano-sized liposomes. Dox loaded versions of these nanoparticles were shown to enhance drug release under induced hyperthermia (43°C). Confocal microscopy and flow cytometry analysis revealed CD44 targeted internalization of liposome nanoparticles in glioblastoma (U87 cells) tumor cells with a dual effect i.e. magneto-thermal and chemotherapeutic triggered the killing of tumor cells *in vitro*, suggesting their potential role as next-generation *in vivo* anti-cancer nano-vehicles [49].

4. Immunogenicity of HA-based nanoparticles

The extent of interaction between nanoparticles and serum proteins, i.e. the formation of the protein corona, is a key factor deciding the intravenous delivery efficiency of the nanoparticles. The protein corona may be modified using coatings that alter the surface properties of the nanoparticles. Almalik et al. showed that modifying chitosan NPs with HA avoided inflammatory protein adsorption compared to nanoparticles not coated with HA [50]. Similarly, reactive oxygen species (ROS) production was suppressed when activated macrophages were treated with HA modified chitosan nanoparticles. The secretion of cytokines such as TNF-α and IL-1β were drastically reduced, indicating low immunogenicity of the HA-CS NPs without any collateral biological responses [51]. Zaki et al. also demonstrated that HA-chitosan NPs were taken up at a relatively slower rate compared to NPs not coated with HA. Moreover, the extent of internalization was two-fold lesser [52]. Further studies are required to verify that HA coatings decrease the immunogenicity of particulates.

5. Conclusions

Research in the development of active cancer-targeting agents led to the discovery of various cell surface molecules that control cellular function via different signaling pathways. CD44 is a cell surface protein that plays a significant role in tumorigenesis, metastasis, and proliferation of cancer cells. Several studies have demonstrated that the interaction of CD44 and HA, an extracellular component, leads to the progression, growth, and metastasis of cancer cells via different signaling pathways. Consequently, strategies have been developed to fabricate HA mediated tumor targeting nanoplatforms. Moreover, it has been reported that conjugation of HA with different nanoparticles increases the internalization of therapeutic molecules via enhanced permeability and retention or CD44 mediated endocytosis with increased therapeutic efficacy both *in vivo* and *in vitro*. Particularly, various drug loaded targeting strategies have emerged, including redox, thermosensitive, and pH sensitive self-assembled HA-prodrug delivery and HA modified metallic and non-metallic nanovehicle-mediated delivery. In summary, HA-based nanoliposomes, micelles, and nano-carriers are promising therapeutic platforms for the delivery of multifunctional cargo in the context of active targeted cancer therapy, paving the way for next-generation clinical cancer therapy.

Acknowledgements

Dr. Nilkamal Pramanik acknowledges DST, SERB- DST (NPDF), Govt. of India, for their financial support.

Conflicts of interest

No conflict of interest is declared.

Author details

Nilkamal Pramanik* and Sameer Kumar Jagirdar
Centre for BioSystems Science and Engineering, Indian Institute of Science, Bengaluru, Karnataka, India

*Address all correspondence to: nilkamalorganic@gmail.com

References

[1] Han HS, Choi KY, KoH, et al. Bioreducible core-crosslinked hyaluronic acid micelle for targeted cancer therapy. J Control Release. 2015; 200:158-166. DOI: 10.1016/j.jconrel.2014.12.032.

[2] Choi KY, Min KH, Yoon HY, et al. PEGylation of hyaluronic acid nanoparticles improves tumor targetability in vivo. Biomaterials. 2011; 32:1880-1889. Doi: 10.1016/j.biomaterials.2010.11.010.

[3] Delmage, JM,Powars, DR, Jaynes, PK, Allerton, SE. The selective suppression of immunogenicity by hyaluronic acid. Ann Clin Lab Sci. 1986; 16(4):303-310.

[4] Prajapati, VD, Maheriya, PM. Hyaluronic acid as potential carrier in biomedical and drug delivery applications. In: Maiti S, Jana S. Editors. Functional Polysaccharides for Biomedical Applications. Ist ed. Woodhead Publishing. Elseviewer, 2019. p. 213-265. Doi.org/10.1016/B978-0-08-102555-0.00007-8.

[5] Huang G, Huang H. Application of hyaluronic acid as carriers in drug delivery. Drug Delivery. 2018; 25: 766-772. Doi: 10.1080/10717544.2018.1450910..

[6] Lee H, Lee K, Park TG. Hyaluronic acid-paclitaxel conjugate micelles: synthesis, characterization, and anti-tumor activity. Bioconjugate Chem. 2008; 19: 1319-1325. Doi: 10.1021/bc8000485.

[7] Klingbeil P, Natrajan R, Everitt G, et al. CD44 is overexpressed in basal-like breast cancers but is not a driver of 11p13 amplification. Breast Cancer Res Treat. 2010; 120: 95-109. Doi.org/10.1007/s10549-009-0380-7.

[8] Lakshman M, Subramaniam V, Rubenthiran U, et al. CD44 promotes resistance to apoptosis in human colon cancer cells. Exp Mol Pathol. 2004; 77: 18-25. DOI: 10.1016/j.yexmp.2004.03.002.

[9] Ranuncolo SM, Ladeda V, Specterman S, et al. CD44 expression in human gliomas. J Surg Oncol. 2002; 79: 30-36. DOI: 10.1002/jso.10045.

[10] Lukyanov, AN, Elbayoumi, TA, Chakilam, AR, Torchilin, VP. Tumor-targeted liposomes: doxorubicin-loaded long-circulating liposomes modified withanti-cancer antibody. J. Controlled Release. 2004; 100: 135-144. DOI: 10.1016/j.jconrel.2004.08.007.

[11] Datir SR, Das M, Singh RP, et al. Hyaluronate tethered,"smart" multiwalled carbon nanotubes for tumor-targeted delivery of doxorubicin. Bioconjugate Chem. 2012; 23: 2201-2213. Doi.org/10.1021/bc300248t.

[12] Yin, T, Liu, J, Zhao, Z, Zhao Y, Dong, L, Yang, M, Zhou J, and Huo, M. Redox Sensitive Hyaluronic Acid-Decorated Graphene Oxide for Photothermally Controlled Tumor-Cytoplasm-Selective Rapid Drug Delivery. Adv. Funct. Mater. 2017; 27: 1604620. Doi.org/10.1002/adfm.201604620.

[13] Cai J, Fu J, Li R, Zhang, F, Ling G, Zhang, P. A potential carrier for anti-tumor targeted delivery-hyaluronic acid Nanoparticles. Carbohydrate Polymers. 2019; 208: 356-364. DOI:10.1016/j.carbpol.2018.12.074.

[14] Meyer K, Palmer JW: The polysaccharide of the vitreous humor. J Biol Chem 1934; 107: 629-634.

[15] Nagabhushan M, Pretlow TG, Guo YJ, et al. Altered expression of CD44 in human prostate cancer during progression. Am J Clin Pathol. 1996; 106: 647-651. Doi.org/10.1093/ajcp/106.5.647.

[16] Wang SJ, Wong G, De Heer AM, et al. CD44 variant isoforms in head and neck squamous cell carcinoma progression. Laryngoscope. 2009; 119: 1518-1530. Doi: 10.1002/lary.20506.

[17] Sugiyama M, Woodman A, Sugino T, et al. Non-invasive detection of bladder cancer by identification of abnormal CD44 proteins in exfoliated cancer cells in urine. Clin Mol Pathol. 1995; 48: 142-147. Doi: 10.1136/mp.48.3.m142.

[18] Ponta H, Sherman L, Herrlich PA. CD44: From adhesion molecules to signalling regulators.Nat. Rev. Cell Biol. 2003; 4: 33-45. Doi.org/10.1038/nrm1004.

[19] Sze JH, Brownlie JC, Love CA. Biotechnological production of hyaluronic acid: a mini review.3Biotech. 2016; 6:67. DOI: 10.1007/s13205-016-0379-9.

[20] Chong BF, BlankLM, Mclaughlin R, Nielsen, LK. Microbial hyaluronic acid production. Applied Microbiology and Biotechnology. 2005; 66, 341-351. DOI:10.1007/s00253-004-1774-4.

[21] Mao Z, Chen RR: Recombinant synthesis of hyaluronan by Agrobacterium sp. Biotechnol Prog 2007; 23:1038-1042. Doi: 10.1021/bp070113n.

[22] Yu HM, Stephanopoulos G: Metabolic engineering of Escherichia coli for biosynthesis of hyaluronic acid. Metabolic Eng 2008; 10:24-32. DOI:10.1016/j.ymben.2007.09.001.

[23] Sheng JZ, Ling PX, Zhu XQ, Guo XP, Zhang TM, He YL, Wang FS: Use of induction promoters to regulate hyaluronan synthase and UDP-glucose-6-dehydrogenase of Streptococcus zooepidemicus expression in Lactococcus lactis: a case study of the regulation mechanism of hyaluronic acid polymer. J Appl Microbiol. 2009; 107: 136-144. Doi. org/10.1111/j.1365-2672.2009.04185.x.

[24] Lu, X, Kamat, MN, Huang, L, Huang, X. Chemical Synthesis of a Hyaluronic Acid Decasaccharide. J Org Chem. 2009; 74(20): 7608-7617. Doi: 10.1021/jo9016925.

[25] Schante, CE, Zubera, G, Herlinb, C, Vandamme, TF. Chemical modifications of hyaluronic acid for the synthesis of derivatives for a broad range of biomedical applications. Carbohydrate Polymers. 2011; 85: 469-489. Doi. org/10.1016/j.carbpol.2011.03.019.

[26] Yin, T, Wang J, Yin L, Shen L, Zhou J, Huo, M. Redox-sensitive hyaluronic acid–paclitaxel conjugates micelles with high physical drug loading for efficient tumor therapy. Polym. Chem. 2015; 6: 8047. DOI:10.1039/C5PY01355K.

[27] Auzenne E, Ghosh SC, Khodadadian M, Rivera B, Farquhar D, Price RE, Ravoori M, Kundra V, Freedman RS, Klostergaard J. Hyaluronic acid-paclitaxel: anti-tumor efficacy against CD44(+) human ovarian carcinoma xenografts. Neoplasia. 2007; 9: 479-486. Doi:10.1593/neo.07229.

[28] Rosato A, Banzato A, Luca GD, Renier D, Bettella F, Pagano C, Esposito G, Zanovello P, Bassi P. HYTAD1-p20: a new paclitaxel-hyaluronic acid hydrosoluble bioconjugate for treatment of superficial bladder cancer. Urol. Oncol.-Semin. Original Invest. 2006; 24: 207-215. Doi.org/10.1016/j.urolonc.2005.08.020.

[29] Liao J, Zheng H, Fei Z, Lu B, Zheng H, Li D, Xiong X, Yi Y. Tumor-targeting and pH-responsive nanoparticles from hyaluronic acid for the enhanced delivery of doxorubicin. International Journal of Biological Macromolecules. 2018; 113: 737-747. DOI:10.1016/j.ijbiomac.2018.03.004.

[30] Li WH, Yi X, Liu X., Zhang Z., Fu Y, & Gong T. Hyaluronic acid ion pairing nanoparticles for targeted tumor therapy. Journal of Controlled Release. 2016; 225: 170-182. Doi.org/10.1016/j.jconrel.2016.01.049.

[31] Ling X, Zhao C, Huang L, Wang Q, Tu J, Shen Y, Sun C. Synthesis and characterization of hyaluronic acid–platinum (IV) nanoconjugate with enhancedantitumor response and reduced adverse effects. RSC Adv. 2015; 5: 81668. Doi.org/10.1039/C5RA16757D.

[32] Li J, MeirongHuo M, Wang J, Zhou J, Mohammad JM, Zhang Y, Zhu Q, Waddad AY, Zhang Q. Redox-sensitive micelles self-assembled from amphiphilic hyaluronic aciddeoxycholic acid conjugates for targeted intracellular delivery of paclitaxel. Biomaterials. 2012; 33: 2310-2320. Doi: 10.1016/j.biomaterials.2011.11.022.

[33] Deng L, Wang G, Ren J, Zhang B, Yan J, Li W, Khashab NM. Enzymatically triggered multifunctional delivery system based on hyaluronic acid micelles. RSC Advances. 2012; 2: 12909-12914. DOI: 10.1039/C2RA21888G.

[34] Liu X, Li W, Chen T, Yang Q, Huang T, Fu Y, Gong T, Zhang Z. Hyaluronic acid modified micelles encapsulating Gem-C12 and HNK for glioblastoma multiforme chemotherapy. DOI: 10.1021/acs.molpharmaceut.7b01035.

[35] Gao S, Wang J,Tian R, Wang G,Zhang L,Li Y,Li L,Ma Q, Zhu L. Construction and Evaluation of a Targeted Hyaluronic Acid Nanoparticle/Photosensitizer Complex for Cancer Photodynamic Therapy. ACS Appl. Mater. Interfaces. 2017; 9(38): 32509-32519. Doi.org/10.1021/acsami.7b09331.

[36] Liu X, Li W,Chen T,Yang Q,Huang T, Fu Y,Gong T,Zhang Z. Hyaluronic Acid-Modified Micelles Encapsulating Gem-C12 and HNKfor Glioblastoma Multiforme Chemotherapy. Mol. Pharmaceutics. 2018; 15: 1203-1214. Doi.org/10.1021/acs.molpharmaceut.7b01035.

[37] Sun Y,Li X,Zhang L, Liu X,Jiang B,Long Z,Jiang Y. Cell Permeable NBD Peptide-Modified Liposomes by Hyaluronic Acid Coating for the Synergistic Targeted Therapy of Metastatic Inflammatory Breast Cancer. Mol. Pharmaceutics. 2019; 16: 1140-1155. DOI: 10.1021/acs.molpharmaceut.8b01123.

[38] FongYT,ChenCH,ChenJP. Intratumoral delivery of doxorubicin on folate-conjugated graphene oxide by in-situ forming thermo-sensitive hydrogel for breast cancer therapy. Nanomaterials. 2017; 7:388-411. Doi: 10.3390/nano7110388.

[39] Xu K, Lee F, Gao SJ, Chung JE, Yano H, Kurisawa M. Injectable hyaluronic acid-tyramine hydrogels incorporating interferon-α2a for liver cancer therapy. Journal of Controlled Release. 2013; 166: 203-210. Doi.org/10.1016/j.jconrel.2013.01.008.

[40] Ashwinkumar N, Maya S, Jayakumar R. Redox-responsive cystamine conjugated chitin–hyaluronic acid composite nanogels. RSC Adv. 2014; 4: 49547. Doi.org/10.1039/C4RA06578F.

[41] Luo Y, Cai X, Li H, Lin Y, Du D. Hyaluronic Acid-Modified Multifunctional Q-Graphene for Targeted Killing of Drug-Resistant Lung Cancer Cells.ACS Applied Materials & Interfaces.2016;8(6): 4048-4055. DOI: 10.1021/acsami.5b11471.

[42] Jung HS, Kong WH, Sung DK, Lee MY, Beack SE, Keum DH, Kim KS, Yun SH, Hahn SK. Nanographene Oxide Hyaluronic Acid Conjugate for Photothermal Ablation Therapy of Skin Cancer. ACS Nano. 2014; 8(1): 260-268. Doi.org/10.1021/nn405383a.

[43] Pramanik N,Ranganathan S, Rao S,Suneet K,Jain S,Rangarajan A,Siddharth Jhunjhunwala, S. A Composite of Hyaluronic Acid-Modified Graphene Oxide and IronOxide Nanoparticles for Targeted Drug Delivery andMagnetothermal Therapy. ACS Omega. 2019; 4(5): 9284-9293. Doi. org/10.1021/acsomega.9b00870.

[44] Yu M, Jambhrunkar S, Thorn P, Chen J. Gu W, Yu C. Hyaluronic acid modified mesoporous silica nanoparticles for targeted drug delivery toCD44 over-expressing cancer cells. Nanoscale. 2013; 5: 178-183. Doi. org/10.1039/C2NR32145A.

[45] Palanikumar L,Kim J,Oh JY, Choi H,Park MP,Kim C, Ryu JH. Hyaluronic Acid-Modified Polymeric Gatekeepers on Biodegradable Mesoporous Silica Nanoparticles for Targeted Cancer Therapy. ACS Biomater. Sci. Eng. 2018; 4(5): 1716-1722. Doi.org/10.1021/ acsbiomaterials.8b00218.

[46] Kang SH; Nafiujjaman M; Nurunnabi M, Li L; Khan HA, Cho KJ, Huh KM; Lee YK. Hybrid photoactive nanomaterial composed of gold nanoparticles, pheophorbide-A and hyaluronic acid as a targeted bimodal phototherapy. *Macromol. Res.* 2015;*23:* 474-484. Doi.org/10.1007/ s13233-015-3061-x.

[47] Wang R, Luo Y, Yang S, Lin J, Gao D, Zhao Y, Liu J, Shi X, Wang X. Hyaluronic acid-modified manganese-chelated dendrimer-entrapped gold nanoparticles for the targeted CT/MR dual-mode imaging of hepatocellular carcinoma. Sci. Rep. 2016;*6:* 1-10. DOI: 10.1038/srep33844.

[48] Luo, Y, Li Y,Li J,Fu C, Yu X, Wu L. Hyaluronic acid-mediated multifunctional iron oxide-based MRI nanoprobes for dynamic monitoring of pancreatic cancer.**RSC Adv.**, 2019;**9:** 10486-10493. https://doi.org/10.1039/ C9RA00730J.

[49] Jose G, LuYJ, Chen HA, Hsu HL, Hung JT, AnilkumarT.S.Chen JP. Hyaluronic acid modified bubble generating magnetic liposomes for targeted delivery of doxorubicin, Journal of Magnetism and Magnetic Materials. 2019; 474: 355-364. Doi. org/10.1016/j.jmmm.2018.11.019.

[50] Almalik A, Benabdelkamel H, Masood A, Alanazi IO, Alradwan I, Majrashi MA, Alfadda AA, Alghamdi WM, Alrabiah H, Tirelli N, Alhasan AH. Hyaluronic Acid Coated Chitosan Nanoparticles Reduced the Immunogenicity of the Formed Protein Corona.Scientific Reports. 2017; 7: 10542. Doi.org/10.1038/ s41598-017-10836-7.

[51] Almalik A, Alradwan I, Majrashi MA, Alsaffar BA, Algarni AT, Alsuabeyl MS, Alrabiah H, Tirelli N, Alhasan AH. Cellular Responses of Hyaluronic Acid Coated-Chitosan Nanoparticles. Toxicol. Res. 2018; **7:** 942-950. DOI: 10.1039/C8TX00041G.

[52] Zaki NM, Nasti A, Tirelli N. Nanocarriers for Cytoplasmic Delivery: Cellular Uptake and Intracellular Fate of Chitosan and Hyaluronic Acid-Coated Chitosan Nanoparticles in a Phagocytic Cell Model. Macromol. Biosci. 2011; 11: 1747-1760. Doi: 10.1002/ mabi.201100156.

Chapter 8

Plant Natural Products: A Promising Source of Hyaluronidase Enzyme Inhibitors

Muhammad Zeeshan Bhatti and Aman Karim

Abstract

Hyaluronidase enzyme degrades hyaluronan, the primary component of the extracellular matrix found in connective tissues animals and on the surface of certain pathogenic bacteria. The degradation of hyaluronan is linked to a wide range of physiological and pathological process. Inhibiting the hyaluronidase enzyme is thus significant as an approach to treat a variety of diseases and health conditions such as anti-fertility, anti-tumor, antimicrobial, and anti-venom/toxin agents. HAase inhibitors of different chemical types have been identified include both synthetic compounds and constituents obtained from naturally sources. Plant natural products as HAase inhibitors are unique due to their structural features and diversity. Medicinal plants have historically been used as contraceptives, antidote for snakebites and to promote wound healing. In recent years, small molecules, particularly plant natural products (alkaloids, flavonoids, polyphenol and flavonoids, triterpenes and steroids) possessing potent HAase have been discovered. A number of plant species from various families, which have folk medicinal claims for these ailments (related to hyaluronan disturbances) were scientifically proven for their potential to block HAase enzymes.

Keywords: hyaluronidase inhibitors, natural products, medicinal plant, phytochemicals

1. Introduction

Hyaluronan/hyaluronic acid (HA) is a biologically important polysaccharide molecule found in the animal kingdom, most notably in the extracellular matrix (ECM) of connective tissues and on the surface of certain pathogenic bacteria. Although HA is found in nearly every tissue of vertebrates, it is abundantly present in the extracellular matrix of soft connective tissues. In mammals, it's predominantly found in the connective tissue of skin, testes, umbilical cord and synovial fluid. HA is composed of a linear polymeric chain with a uniform repeating disaccharide units of D-glucuronic acid and N-acetyl-D-glucosamine linked through (1,3 and 1,4) glycosidic bond. HA is a megaDalton molecule, synthesized as free polymer by the plasma membrane at its inner face [1–3].

The molecular function of hyaluronan in the body include interaction with HA receptors on the surface of the same cell or ECM molecules of the surrounding cells [4, 5]. When newly secreted, HA interacts with a variety of cell surface receptors

(CSRs) which give rise to important physiological functions such as signal transduction, building of pericellular matrix and the degradation endocytosis of HA via receptor-mediated internalization [6–8].

The metabolism of hyaluronan involves hyaluronidase enzyme, which is a class of glycosidase that predominantly degrades hyaluronan (HA). Karl Meyer coined the word Hyaluronidases (HAases), and over the years of research, the importance of HAases in controlling the physiological and pathological function of HA in animals has been established [9]. In mammals, the HAases hydrolyze the glucosaminidic β-1,4-linkages of hyaluronic acid and produces tetrasaccharide fragments. Three types of HAase enzymes act in concert to degrade HA biochemically; first, intact HA is acted on by endoglycosidase HAases, resulting in oligosaccharides with varying chain lengths that serve as substrates for the other two HAase enzymes (exoglycosidases), namely -glucuronidase and -N-acetyl hexos [10, 11].

The enzyme hyaluronidase and its substrate (Hyaluronan) perform a critical biological function in human body and their imbalance has been linked with various pathological processes and disease states including skin diseases and cancer [1]. The biological role of HA depends on the type of product formed after degradation and the circumstances under which it is synthesized [6, 12]. The involvement of HA has been established in various physiological and pathological processes include embryogenesis [7, 13, 14], immune surveillance, inflammation [15–18], wound healing [19], multi-drug resistance [6], cancer, water homeostasis and viscoelasticity of ECM [6, 8, 11, 20, 21]. Thus, it is critical to maintain HA homeostasis by balancing the action of HAase enzymes involved in anabolic and catabolic activities using various approaches such as hyaluronidase enzyme inhibitors (HAIs). The biological and therapeutic potential of HAase inhibitors (HAIs) is receiving significant attention, and an increasing amount of research is being conducted to develop potent hyaluronidase inhibitors for a variety of health conditions, including contraceptives, anti-tumor, antimicrobial, and anti-venom/toxin agents [22–24]. Hyaluronidase inhibitors of different chemical types are increasingly being reported, which include synthetic and plant derived bioactive compounds, polysaccharides, fatty acids, proteins, glycosaminoglycans and others [23, 25–30].

In this chapter, we have discussed and presented an updated overview of studies on important natural product agents (small molecules, and plants extracts) of various chemical forms derived from medicinal plants, which have been reported as potent hyaluronidase inhibitors. The search engines, such as, Google Scholar and PubMed were used to search the literature using key words such as natural products, medicinal plants, phytochemicals with hyaluronidase inhibitors, and anti-hyaluronidase. Majority of the data covered in this study are research published during the last fifteen years and studies with incomplete data or doubtful peer review system were excluded.

2. Hyaluronidases

Hyaluronidases are a family of endoglycosidase enzymes found in both eukaryotes and prokaryotes and prevalent across the animal kingdom [31]. It was first observed by Duran-Reynals in mammalian testis extract and termed it "spreading factor" as it has the property of breaking down the hyaluronan structure and facilitating tissue permeability and spreading [32]. Karl Meyer later classified hyaluronidases into three groups depending on chemical analysis and end products formed, which included mammalian, leech, and bacterial hyaluronidases.

Mammalian hyaluronidases are endo-β-Nacetlyhexosaminidases which arbitrary cleave hyaluronan glycosidic ate β-1-4 position, yielding even numbered tetra- and hexa oligosaccharides as the major end products along with N-acetylglucosamine at the reducing end of the product. These hyaluronidases exhibit both hydrolytic and transglycosidase activity and are found in spermatozoa, mammalian cell lysosomes, and bee, snake, and reptile venoms [33].

The second type of HAases are leech hyaluronidase, which cleave glucoronate linkages of hyaluronan and are inert towards other glycosaminoglycans. These group of HAases are hyaluronate-3-glycanohydrolases are endo-β-D-glucuronidases. Tetra- and hexasaccharides are the main end products with glucuronic acid at the reducing end. This group of enzymes are present in salivary glands of leeches and hook worms [34].

The third type is microbial hyaluronidases, which are distinguished from mammalian and leech HAases by their lack of hydrolysis activity. These HAases catalyze the cleavage of HA at the 1–4 glycosidic bond, resulting in the formation of 4 and 5 member unsaturated oligosaccharides. Enzymes in this class includes HA lyases from *Streptococcus pneumoniae* (S. PHL) and *S. agalactiae* [34, 35].

In humans, six hyaluronidase-like genes known as hyaluronoglucosaminidases (Hyals1–6) have been identified. Of the six Hyal genes, Hyal1 and 2 are the primary hyaluronidases responsible for the catabolism of HA in somatic tissue, while Hyals3 to 6 are inactive and likely do not participate in HA cleavage [36]. Although inactive, hyal3 is widely expressed in chondrocytes, testis, and bone marrow, and its expression increases when fibroblasts differentiate into chondrocytes. Inflammatory cytokines such as IL-1 and TNF- (tumor necrosis factor-alpha) upregulate the Hyal2 and Hyal3 genes, but not the Hyal1 gene [37].

3. Plant derived natural products as hyaluronidase inhibitors

In the regulation of biological processes, inhibition of enzyme activity can be as essential as the activity itself. Many diseases are caused by overactivation of enzymes, which can be regulated with enzyme inhibitors since blocking the enzyme is more efficient in active catabolic reactions than stimulating the synthesis of substrates such as the high molecular weight polymeric hyaluronan contained in the extracellular matrix [38]. This is particularly true when a rapid response or finely regulated temporal and spatial ECM activities are required. Mio and his colleagues have identified the first inhibitor of the hyaluronidase enzyme in human and mouse serum [39].

For centuries, nature has been a source of medicinal products, with numerous useful medicines have been derived from plant sources [40]. Their therapeutic utility in treating a variety of illnesses have been investigated in various conventional medical systems, and their role as a biological modulator has been recognized throughout human history [41]. Natural products' effectiveness as enzyme inhibitors is attributed to their product biosynthetically in living organisms, which enhances their chances of interacting effectively with a variety of biological targets [42]. The inherent steric complexity, more number of rings and chiral centers, as well as the presence of more oxygen and the ability to form more hydrogen bonds, increases drug-likeness property of natural products from synthetic ones [43, 44]. The following section discusses recent research on various plant extracts and phytoconstituents as potential sources of hyaluronidase inhibitors.

3.1 Anti-hyaluronidase phytoconstituents

Various class of natural products derived from different plants species documented as hyaluronidase inhibitors include alkaloids, flavonoids, polyphenols, terpenes and steroids as shown in the **Table 1**. Natural products derived from plants

Class of Natural Products	Compounds	Source of HAase enzyme	IC_{50}/%Inhibition	Ref.
Alkaloids	Aristolochic acid	*Naja naja* venom	50 µM	[28]
	Ajmaline		-	
	Reserpine		-	
	Nuciferine	Testicular	>100 µM	[45]
	Nornuciferine		22.5 µM	
	N-methylasimilobine		>100 µM	
	Asimilobine		11.7 µM	
	Pronuciferine		>100 µM	
	Armepavine		>100 µM	
	Norarmepavine		26.4 µM	
	N-methylcoclaurine		>100 µM	
	Coclaurine		11.4 µM	
	Norjuziphine		24.3 µM	
	Aristolocic acid	Naja naja venom	1.43 µM	[46]
	3-[(4-methylpiperazin-1-yl)methyl]-5-phenyl-1H-indole	Testicular	23% (5 µM)	[47]
Flavonoids/ polyphenols	Flavone Tannic acid Quercetin	Naja naja venom	50 µM	[28]
	Tannin	Honey bee, scorpion, snakes and cobra venoms	0.9%	[53]
	Kaempferol		21.0%	
	Silybin		24.8%	
	Myriceetin		31.5%	
	Morin		33.0%	
	Quercetin		33.9%	
	Butein		37.8%	
	Phloretin		41.1%	
	Catechin		42.5%	
	Flavone		46.9%	
	Rutin		40.5%	
	Isoquercitin		42.1%	
	Apigenin		15.0%	
	Apigenin Kaempferol Luteolin Tannic acid	Honey bee, scorpion, snakes and cobra venoms		[54]

Class of Natural Products	Compounds	Source of HAase enzyme	IC_{50}/%Inhibition	Ref.
	quercetin 3-O-β-D-glucopyranoside	Bovine testes	20.9 mM	[57]
	quercetin 3-O-β-D-xylopyranoside		22.1 mM	
	kaempferol 3-O-β-D-glucopyranoside		26.5 mM	
	isorhamnetin		55.4 mM	
	Rosmarinic acid	Testicular	309 μg/mL	[57]
	Lithospermic acid B		164 μg/mL	
	Diometin-7-O-β-D-glucopyraanoside		644 μg/mL	
	Apigenin-7- O-β-D-glucuronopyranoside		548 μg/mL	
	Tannic acid	Testicular	4.97 units/mL	[58]
	Apigenin		4.02 units/mL	
	Quercentin		4.28 units/mL	
	Tannic acid	Testicular	0.8 units/mL	[59]
	Gallic acid		5 units/mL	
	Ellagic Acid		4.8 units/mL	
	Chicoric acid	*Escherichia coli* F470	171 μM	[68]
Terpenes/ steroids	Glycyrrhizin	*Streptococcus agalactiae*	0.020–1.300 mM	[52]
	Glycyrrhetinic acid	Bovine testes	0.060–0.260 mM	
	3β-urs-12-en-28-oic acid	Testicular	103.18 ±1.70 μM	[62]
	3β,19,23-trihydroxyurs-12-en-28-oic acid		286.95±10.28 μM	
	3β-acetylolean-12-en-28-oic acid triterpenoid		1466.5± 2.37 μM	
	Steroidal Fraction	*Brevibacterium halotolerans* DC1	5.19 mM	[67]
	Testosterone Propionate	*Escherichia coli* F470	124 ± 1.1 μM	[68]
	Glycyrrhizic acid	*Escherichia coli* F470	175 ± 1.2 μM	[68]

Table 1.
Natural product compounds active against hyaluronidase enzyme.

are well-known as HAase inhibitors due to their unique structural features. As indicated in **Table 1**, many classes of natural compounds produced from various plant species have been recorded as hyaluronidase inhibitors. These classes include alkaloids, flavonoids, polyphenols, terpenes, and steroids.

3.1.1 Alkaloids

Alkaloids are naturally occurring secondary metabolites, which consist of a basic nitrogen atom and produced by various species of animals, plants, bacteria and fungi. Morikawa and his team evaluated aporphine and benzylisoquinoline alkaloids which they have earlier isolated from the flower buds of Sacred lotus (*Nelumbo nucifera*) tably, Among the alkaloids discovered as a hyaluronidase inhibitor, nornuciferine (IC_{50} = 22.5 μM), asimilobine (11.7 μM), norarmepavine (26.4 μM), coclaurine (11.4 μM), and norjuziphine (24.3 μM) have shown potent activity, even higher than the standard atilllergic drug (disodium cromoglycate (IC_{50} = 64.8 μM). Nuciferine, N-methylasimilobine, pronuciferine, armepavine, and N-methylcoclaurine are the other alkaloids with moderate anti-HAase action [45]. Girish and co-researchers tested various compounds including well known alkaloids such as aristolocic acid, reserpine, and ajmaline on hyaluronidase enzymes obtained from the *Naja naja* snake venome and observed a dose dependent inhibition of hyaluronidase enzyme activity in manner. It was further observed that aristolochic acid has completely inhibited HAase, while reserpine and ajmaline inhibited it partially in a non-competitive manner. [28, 46]. Olgen and colleagues have tested as a series of aminomethyl indole alkaloids derivatives against the bovine testes hyaluronidase and found 3-[(4-methylpiperazin-1-yl)methyl]-5-phenyl-1H-indole as the most potent inhibitor of HAase enzyme [47].

3.1.2 Flavonoids and polyphenols

Flavonoids are a large group of polyphenolic compounds having benzo-γ-pyrone structure and are ubiquitously present in various parts of the plants. Flavonoids are a wide class of polyphenolic chemicals with a benzo—pyrone structure that are found in virtually every part of plants. Secondary metabolites of phenolic origin, such as flavonoids, are involved in a variety of pharmacological activities [28]. Based on their structure, flavonoids of different types such as flavones, anthocyanidines, flavones, and chalcones have demonstrated antioxidant, anti-inflammatory, antiviral, and antithrombotic properties, antitumor, hepatoprotective and enzyme inhibitory properties [48–50].

Girish and co-researchers have observed in an *in vitro* study that flavonoids of different structure types such as flavone, tannic acid quercetin were able to inhibit the hyaluronidase enzyme activity obtained from *Naja naja* snake venom [28].

In an early study, Rodney and co-researchers evaluated the effect of flavonoids on hyaluronidase and afterwards the effect of 31 flavonoids has been found potent against the activity of bovine testicular hyaluronidase. The inhibitory action of flavonoids on hyaluronidases is dependent on the number of hydroxyl groups and side chain substituents present in the molecules, and flavonoids containing many hydroxyl groups were found to significantly reduce inhibitory activity hyaluronidase enzyme [51]. Plant based flavonoids such as flavones, 2-hydroxy-flavone, apigenin, luteolin, quercetin, and myricetin demonstrated the inhibitory effects on hyaluronidase activity [51].

Herte et al. investigated the effects of several flavonoids on the microbial origin of the hyaluronidase enzyme (Hyaluronate lyases). During their research, they discovered quercetin and myricetin to be the most potent inhibitors, with extra hydroxyl groups at positions 3,3′ (quercetin) and 5′ (myricetin) (myricetin). In addition, glycosylated flavonoids such as rutin, apiin, and silybin have shown a decline in their capacity to inhibit hyaluronate lyase, even when the side groups carried hydroxyl groups themselves [52].

A series of flavonoids were examined by Kuppusamy et al. against hyaluronidase enzyme extracted from the venom of honey bee, scorpion and cobra and found flavonoids such as myricetin, quercetin, luteolin, apigenin, phloretin and kaempferol showing potent anti-HAase effects in *in vitro* assay [53]. In another study the same authors observed a contrast where, sylibin inhibited hyaluronidase activity of bee and scorpion venom supported by the apigenin, kaempferol, luteolin, and tannic acid [54].

Kim and his co-worker isolated flavonols (quercetin 3-O-β-D-glucopyranoside, quercetin 3-O-β-D-xylopyranoside, kaempferol 3-O-β-D-glucopyranoside, and isorhamnetin 3-O-β-D-glucopyranoside) from the *Allium sativum* L. for anti-hyaluronidase properties [55].

Polyhenols are naturally occurring secondary metabolites, largely found in plants and generally involves in the defense of plants against pathogens [56]. Other type of phenolic compounds includes rosmarinic acid, lithospermic acid B, diometin-7-O-β-D-glucopyraanoside, and apigenin-7-O-β-D-glucuronopyranoside reported from the *Meehania fargesii* plant, all of which were effective at suppress HAase activity [57]. Tatemoto and colleagues investigated the effects of tannic acid, apigenin, and quercetin, in *in vitro* fertilization parameters, on hyaluronidase activity, and found that tannic acid was the most active hyaluronidase enzyme activity in a dose-dependent manner at concentrations ranging from 2 to 10 g/ml [58]. The same group of researchers investigated three tannins, tannic acid, gallic acid, and ellagic acid, and found tannic and ellagic acid as potent inhibitors of the hyaluronidase enzyme, effectively preventing polyspermy by suppressing the acrosome reaction induced by sperm-zona interaction during in vitro fertilization of porcine oocytes [59].

3.1.3 Terpenes and steroids

Terpenes are the constituents of pheromones, anti-feedants and flavors, which are composed of isoprene unite (C_5) and their derivatives. Terpenes and terpenoids (oxygenated derivative) are recognized as one of the important class of natural products, are widely distributed in plants and possesses a range of bioactivity, exhibiting a wide bioactivity, such as anticancer, neuroprotection, and anti-inflammation and anti-infective agents [60, 61]. Abdullah and co-authors isolated teriterpenes as HAase blocking agents from *Prismatomeris tetrandra* (Roxb.) K. Schum. The two triterpenoids (3β-urs-12-en-28-oic acid and 3β,19,23-trihydroxyurs-12-en-28-oic acid) were obtained from the chloroform fraction whereas another triterpenoid 3β-acetylolean-12-en-28-oic acid was isolated from the roots of *Prismatomeris tetrandra*. Also, the synthetic analogues of ursolic acid were identified as potential inhibitor of hyaluronidase [62]. The *in-vitro* inhibition of bovine hyaluronidase and hylaluronate lyase was shown by the triterpenes glycyrrhizin and glycyrhetinic acid [52]. However, fatty acid derivative of glycyrrhetinic acid, known as stearyl ester was unable to inhibit the hyaluronidase activity [63]. This difference may be due to the specific structures of the respective enzymes as well as the splitting mechanism of hyaluronic acid as endoenzyme or exoenzyme [64].

Sterols are important structural components in higher organisms. They take part in the regulation of membrane fluidity, permeability and membrane associated metabolic processes [65]. Steroids of different structure types are reported to influence hyaluronidase metabolism [66]. Patil and co-researchers found the steroidal fraction isolated from the leave of *Carissa carandas* as strong inhibitior of hyaluronidase enzyme activity with IC_{50} = 5.19 mM/mL as compared to the standard (quercetin). This steroidal fraction could contain potential hyaluronidase inhibitor and therefore should be considered for further studied as anti-venom agent [67].

In a study, Lengers and team found chicoric acid (IC_{50} = 171 μM) and testosterone propionate (IC_{50} = 124 ± 1.1 μM) as strong inhibitors of Hyal1 expressed over the surface of *Escherichia coli* F470 which was comparable to that of glycyrrhizic acid (IC_{50} = 177 μM) [68].

3.2 Anti-hyaluronidase medicinal plant extracts

Plants have remained a major source of medicine for centuries and therapeutic agents derived from natural sources are used traditionally to recover from wound healing, treat snakebites or inflammation as contraceptives. Several studies indicate that plants species from various families, which have folk medicinal claims for these ailments were also scientifically been proven for their potential to block HAase enzymes as shown in **Table 2**.

Plant Name	Plant part/type of extract (active extract)	Biological activity	Source of HAase enzyme	Ref.
Meehania fargesii (Lamiaceae)	Whole plants 80% acetone extract (water soluble fractions)	Anti-HAase	Testicular	[57]
Aesculus hippocastanum (Hippocastanaceae)	Seeds/ aqueous-ethanol	Anti-inflammatory	Testicular	[69]
Hedera helix (Araliaceae)	Leaf/ aqueous-ethanol	Anti-inflammatory	Testicular	[69]
Hygrophila schulli	Leaf/ethanolic extract	Anti-inflammatory	Testicular	[70]
Areca catechu (Arecaceae)	Whole plant/ aqueous methanol	Anti-aging	Testicular	[71]
Dryopteris cassirrhizoma (Dryopteridaceae)	Methanol–water extract of whole plant	Anti-aging/ Anti-inflammatory	Testicular	[71]
Alpinia katsumadai (Zingiberaceae)	Whole plant/ aqueous-ethanol	Anti-aging/ Anti-inflammatory	Testicular	[72]
Cinnamonum cassia (Lauraceae)	Whole plant/ aqeuous-methanol	Anti-aging/ Anti-inflammatory	Testicular	[72]
Curcuma longa (Zingiberaceae)	Methanol–water extract of whole plant	Anti-aging/ Anti-inflammatory	Testicular	[72]
Prunus salicina (Rosaceae)	Root bark/aqueous decoction	Anti-HAase	—	[73]
Anemarrhena asphodeloides (Asphodelaceae)	Rhizom/methanol extract	Anti-inflammatory/ Anti-allergy	Testicular	[74]
Rubus fruticosus/Blackberry (Rosaceae)	Fruits/Methanol	Anti-Inflammatory	—	[75]
Artocarpus altilis (Moraceae)	Bark/ethanol	Anti-aging	Testicular	[76]
Curcuma aromatica (Zingiberaceae)	Rhizomes/ethanol	Anti-aging	Testicular	[76]

Plant Name	Plant part/type of extract (active extract)	Biological activity	Source of HAase enzyme	Ref.
Chamaerhodos altaica (Rosaceae)	Aerial parts/80% acetone extract (aqueous fraction, BuOH)	Anti-HAase	—	[77]
Camellia sinensis (Theaceae)	Leaves, buds/(water brew)	Anti-aging/skin care	Testicular	[78]
Canavalia gladiata White Sword Beans (Fabaceae)	Seeds/80% methanol extracts(fermented and non-fermented)	Anti-inflammatory	—	[79]
Glycine max Soybeans (Fabaceae)	Seeds/80% methanol extracts (fermented and non-fermented)	Anti-inflammatory	—	[79]
Coffee Rubiaceae (Rubiaceae)	Seeds/Coffee silverskin (byproduct of the roasting procedure for coffee beans	Anti-inflammatory/ Anti-allergy	Testicular	[80]
Deutzia coreana (Hydrangeaceae)	Stem/methanolic extract	Anti-inflammatory/ Anti-allergy	Testicular	[81]
Osmanthus insularis (Oleaceae)	Stem/methanol	Anti-inflammatory/ Anti-allergy	Testicular	[81]
Styrax japonica (Styracaceae)	Stem/methanol extract	Anti-inflammatory/ Anti-allergy	Testicular	[81]
Dracocephalum foetidum (Lamiaceae)	Aerial parts/80% aqueous acetone extract (aqueous fraction)	Anti-inflammatory	Testicular	[82]
Keiskea japonica (Lamiaceae)	Aerial part/80% acetone extract	Anti-HAase	Testicular	[87]
Lycopus lucidus (Lamiaceae)	Aerial part 80% acetone extract	Anti-HAase	Testicular	[88]
Lythrum salicaria L (Lythraceae)	Whole Plant/ aqueous extract	Anti-inflammatory	Testicular	[89]
Terminalia chebula (Combretaceae)	Fruit dried/95% ethanol extract	Anti-fertility	Human spermatozoa	[90]
Gaultheria procumbens (eastern teaberry) (Ericaeae)	Leaves/Petroleum ether, chloroform	Anti-inflammatory	Testicular	[91]
Payena dasyphylla (Sapotaceae)	Bark/methanolic extract	Anti-arthritic	Testicular	[92]
Phyllanthus emblica (Phyllanthaceae)	Aqueous extract of fruit	Chondroprotective	Testicular	[93]
Vitis rotundifolia (Vitaceae)	Seed and skin/50% ethanol extract	Anti-HAase	Testicular	[94]
Malaxis acuminata (Orchidaceae)	Leaves, stem/ methanolic extract	Skin-aging	—	[95]
Mimosa pudica (Fabaceae)	Roots/aqueous extract	Anti-ophidian	Snake venom	[96]

Plant Name	Plant part/type of extract (active extract)	Biological activity	Source of HAase enzyme	Ref.
Oenothera biennis Evening-primrose (Onagraceae)	Aerial Part/50% methanolic extract	Anti-inflammatory	—	[97]
Oenothera paradoxa Evening-primrose (Onagraceae)	Aerial Part/50% methanolic extract	anti-inflammatory	—	[97]
Otostegia fruticosa (Lamiaceae)	Leaf/70% ethanolic extract	Anti-inflammatory	Testicular	[98]
Brown algae (*Eisenia bicyclis* and *E. kurome*)	Crude phlorotaninin extract	Anti-aging	Testicular	[99]
Padina pavonica (Dictyotaceae)	Seaweed/extracts (Pressurized liquid extraction, microwave assisted extraction. Supercritical fluid extraction)	Anti-aging	Testicular	[100]

Table 2.
Medicinal plants with hyaluronidase activity.

In a bioassay directed study, the polar fraction of *Aesculus hippocastanum* L (seeds) and *Hedera helix* L (leaves) were found active against hyaluronidase enzyme and later isolation of triterpene and steroidal saponins and sapogenins also exhibited strong anti-HAase activity when tested against testicular hyaluronidase enzyme [69].

The well-known medicinal plant *Hygrophila schulli*, traditional used as anti-inflammatory and pain treatment in the north Ethiopia and India, possess anti-hyaluronidase activity *in vivo* by the ethanolic leaf extract [70]. In a study by Lee et al. [71], the aqueous methanolic extracts of 150 plant species assayed for their potential as hyaluronidase inhibitors, the extracts of six species found to be most active against HAases were *Areca catechu*, *Alpinia katsumadai*, *Dryopteris cassirrhizoma*, *Cinnamonum cassia*, and *Curcuma longa*, and the extract of *Areca catechu* showed relatively higher anti-HAase activity. The major constituents identified in *Areca catechu* which include phenolic compounds such as flavonoids and tannins could be responsible for the anti-HAase effect [72].

In another anti-HAase screening study, Tomohara et al. [73] evaluated the decoction extracts of 98 plant species for HAase inhibitory activity in an *in vitro* HAase assay. They observed 17 extracts exhibited moderate to high inhibitory activity (>50% inhibition at 500 μg/mL) and also noted correlation between the total phenol present in the extract and their cumulative effect as anti-HAase activity was varying. From the study, rhizome extract of *Panax japonicus* and root bark extract of *Prunus salicina* were found most potent HAase inhibitors.

Jeong et al. [74] evaluated 100 Korean medicinal plants for their anti-allergic activity. The methanolic rhizome extract of *Anemarrhena asphodeloides* was found active against hyaluronidase enzyme. Marquina and colleagues investigated the anti-inflammatory effect of blackberry fruit extract and fractions. Two of the seven fractions inhibited the hyaluronidase enzyme and shown superior anti-inflammatory effect when compared to aspirin [75].

A study conducted by Liyanaarachchi et al. [76] on fifteen Sri Lankan medicinal plants for their skin aging and anti-wrinkle effect has identified three plant extract with relatively higher anti-HAase activity. The ethanol extract of *Curcuma aromatica* rhizomes exhibited marked hyaluronidase inhibitory activities (95.0%) inhibition at 500 μg/mL) followed by *Artocarpus altilis* (68.59%) and *Artocarpus nobilis* bark extracts (44.78%) when tested at 500 μg/mL concentration level.

Selenge et al. [77] studied two medicinal plant famous in Mongolian traditional medicine *Chamaerhodos erecta* and *C. altaica* and revealed in the bioassay guided isolation the moderate ability of its constituents as anti-HAase enzyme, suggesting their potential to prevent the extracellular matrix degradation factors.

Similarly, the Sri Lankan origin black tea *Camellia sinensis* L. (Orthodox Orange Pekoe) was evaluated for its potential as cosmeceutical for skin aging by Ratnasooriya et al. [78]. The extract revealed moderate anti-HAase activity (IC_{50} =1.09 ±0.12 mg/mL) compared to standard compound (epigallocatechin gallate) (IC_{50} = 0.09±0.00 mg/mL) in as dose dependent manner.

Han et al. [79] assayed the fermented and non-fermented seed's methanolic seed extract of White Sword Beans (*Canavalia gladiata* DC) and Soybeans (*Glycine max* L. Merrill) and found higher inhibitory activity exhibited by red sword beans (non-fermented/fermented) (1.5–2.6-fold) against HAase enzyme than that of soybeans (non-fermented/fermented). The study suggests that *B. subtilis*-fermented sword beans are potential as potential anti-inflammatory agents for the food industry.

Furusawa et al. [80] investigated the silverskin coffee beans (a by-product during roasting) for its anti-inflammatory and anti-allergic effects. The results indicated a potent inhibitory effect against hyaluronidase (IC_{50}=0.27 ±0.04 mg/mL) as compared to the standard disodium cromoglycate (IC_{50} = 0.31±0.05 mg/mL). The strong effect is argued possibly due to the presence of acidic polysaccharides present in the extract, which is mainly composed of uronic acid present in Silverskin coffee beans extract.

A major screening study on 500 Korean Medicinal plants as HAase inhibitors identified the stem extract of three species possessing relatively higher anti-HAase activity include plant specied *Styrax japonica* (57.28%), *Deutzia coreana* (53.50%), and *Osmanthus insularis* (53.19%). The study further explores that the HAase inhibition could be due to presence of multifunctional compounds and may be effective in preventing allergic reactions and inflammation [81].

Dracocephalum foetidum Bunge, is a medicinal plant traditionally used by Mongolian nomads for various infections and suppurative disease and fever. Its chemical and physiological role was investigated by Selenge et al. [82] and found the aqueous acetone extract of the aerial part possess potent anti-hyaluronidase activity Acetone extract (IC_{50} = 0.27 ± 0.01 mg/mL) compared to the standard (disodium cromoglicate, IC_{50} = 0.33 ± 0.02 mg/mL) and further phytochemical analysis of its aqueous fraction resulted into compounds of various class as highly potent HAase inhibitors.

Załuskia et al. [83] found strongest inhibitory effects in the freshly dried fruits of *Eleutherococcus senticosus* (IC_{50} = 0.58 ± 0.01 mg/mL) and *E. henryi* (IC_{50} = 0.61 ± 0.05 mg/mL), compared to positive control (Methyl indole-3-carboxylate, (IC_{50} = 07.11 mM). *Eleutherococcus senticosus* Maxim. called as Siberian ginseng, and has been used from ancient times in Northeastern Asia and Eastern Russia as a tonic and anti-fatigue agent. In northeast China, the ethanolic roots extract is a popular health supplement for weakness, rheumatism, impotence and hemorrhoids [84, 85]. *E. senticosus* products are imported in Europe and it is one of the ten popular herbal dietary supplements in North America [86].

Murata and co-researchers investigated three the 80% acetone extracts of three medicinal plants, *Keiskea japonica, Lycopus lucidusand* and *Meehania fargesii* for their inhibitory effect against hyaluronidase. From bioassay directed study of these plants, the phenylpropanoids and flavone glucuronide from aerial part of *Keiskea japonica* [87], phenylpropanoids from aerial parts *Lycopus lucidas* [88] and spermidine alkaloids flavone glycosides from dried whole plant extract of *Meehania fargesii* [57] showed strong hyaluronidase inhibitory activity.

Piwowarski and group examined tannin-rich aqeous extract of twelve plant for their ability to inhibit hyaluronidase materials based on their use in traditional Polish medicine for external treatment of skin and mucosal diseases. Among the plants, *Lythrum salicaria* L. extract has shown strongest inhibition of hyaluronidase (IC_{50} = 8.1 ± 0.8 μg/mL) compared to the heparin (IC_{50} = 62.1 ± 7.5 μg/mL) which was used as standard control [89].

Terminalia chebula, an Indian medicinal plant was assessed for its role as male antifertility agent using hyaluronidase inhibition enzyme assay. The 95% ethanol dried fruit extract of the plant showed in vitro HAase inhibitory activity of the human spermatozoa (~93% inhibition,) (IC_{50} = 0.8579 mg/ml) and rat caudal epididymal spermatozoa (~86% inhibition) (IC_{50} = 1.6221 mg/ml) at 30 mg/mL compared to the standard tannic acid (IC_{50} =299.6 μM). In the *in vivo* study on rates showed statistically significant ($P < 0.001$) inhibition of hyaluronidase activity of HAase enzyme extracted from testis (50 mg/kg dose, −47% decrease) and caused a further decrease (−72% decrease) at 100 mg/kg dose. The anti-HAase activity of the extract against caput and cauda epididymal spermatozoa extracted enzyme exhibited significantly better ($P < 0.001$) activity at 50 mg/kg dose (−41% each) and 100 mg/kg dose (−65% and −77%, respectively) when given orally for 60 days [90].

Michel and colleagues investigated the anti-inflammatory properties of Eastern Theaberry (*Gaultheria procumbens*) [91], found the chloroform (IC_{50} = 282.15 ± 10.38 μg/mL) and pet-ether (IC_{50} = 401.82 ± 16.12 μg/mL) fractions of the plant leaf extract as potent hyaluronidase inhibitors compared to the standard drug heparin (IC_{50} = 366.24 ± 14.72 μg/mL) which was higher than the activity they observed in nine most active constituents present in the sample.

Citalingam and Co-researchers have screened different extracts prepared from the bark and leaves of *Payena dasyphylla* medicinal plants for their potential as anti-hyaluronidase inhibitors. It was found to exert higher activity (IC_{50} = 100 μg/mL) against bovine testicular hyaluronidase The *Payena dasyphylla* extract also showed strong inhibition of HAase expression (IC_{50} = 100 ng/mL) in the cultured human chondrocyte cells in response to IL-1β. Similarly, the ethyl acetate fraction of the plant has strongly exhibited inhibited the HYAL1 and HYAL2 mRNA gene expressions (IC_{50} = 100 μg/mL) [92].

Phyllanthus emblica is a rejuvenating plant famous in Ayurvedic medicine has been evaluated by Sumantran et al. for its ant-arthritic property [93]. The aqueous decoction extract fruit (powder) was found to inhibit the activity of HAase enzyme effectively at IC_{50} = 0.15 mg/ml.

The muscadine grape (*Vitis rotundifolia*) is scientifically known for its anti-inflammatory properties. Bralley and co-authors have tested the ethanol extract of fruit skin and seed for their inhibitory potential against hyaluronidase enzyme. In their study they observed the bronz IC_{50} = 0.3 mg/mL for) and purple IC_{50} = 0.6, mg/mL) mascadine seed extracts as highly potent compared to the fruit skin extracts of the two mascadine types (IC_{50} = 1.0, 1.0 mg/mL respectively) [94].

Malaxis acuminata, an important medicinal plant known in Ayurvedic medicine was evaluated for its effect on skin aging and related enzyme activity. The researchers found the in vitro- isolated leaf extract (Methanolic) as strong inhibitor of hyaluronidase activity (IC_{50} = 60.36 ± 1.6 μg/mL) compared to the standard compound oleanolic acid-known for skin protective effect (IC_{50} = 32.45 ± 1.7 μg/mL) [95].

Girish et al. demonstrated that the aqueous root extract of *Mimosa pudica* reduced the hyaluronidase activity of three Indian snake venoms; *Naja naja* (2.16×10^{-3} ± 0.04), *Vipera russelii* (1.25×10^{-3} ± 0.045) and *Echis carinatus* (1.1×10^{-3} ± 0.072) units/min/mg protein [96].

Plants bearing high amount of tannin are known to block Hyaluronidase enzyme. A study conducted by Granica et al. discovered the extracts made from aerial part of two plants Oenothera paradoxa Hudziok and *O. biennis* L, which are rich in macrocyclic ellagitannin showed strong inhibition (97.3 ± 3.0% and 97.9 ± 1.7% respectively) of the HAase enzyme activity at 50 μg/mL compared to the stander heparine (62.1 ± 7.5 μg/mL) [97].

In a recent study on *Otostegia fruticose*, a medicinal plant traditionally used in Ethiopia to treat different ailments including inflammatory disorders. In the study, Bahta and co-researchers [98] found the ethanolic leaf extract and its fractions a dose depended anti-HAase activity. The crude ethanolic extract and chloroform fraction exhibited highest hyaluronidase inhibition (79.20% and 85.75% respectively), compared to standard drug indomethacin (95.52%) at the concentration of 100 μg/mL.

Brown algae are a nutrient-dense and potential source of bioactive secondary metabolites. In a study conducted by Shibata and co-workers [99], the crude phlorotaninin extract of two brown algae (*Eisenia bicyclis* and *E. kurome*) exhibited stronger anti-HAase activity (IC_{50} = 0.03 and 0.035 mg/mL respectively) compared to the two standard compounds Epigallocatechin gallate (IC_{50} = 190 mM/mL and Sodium cromoglycate (IC_{50} = 270 mM/mL). In the same study they observed its constituents possessing strong inhibitory activity.

In another study on algae, Fayad and co-researcher have used capillary electrophoresis-based enzymatic assay method to assess the anti-skin aging property of a macroalga (*Padina pavonica*). In their study, the water extract was found strongly inhibiting the hyaluronidase activity (IC_{50} = 0.04 ± 0.01 mg/mL) compared to the literature reported value of phlorotannin fractions of *Einsenia bicyclis* (IC_{50} = 0.03 mg/mL) [100].

4. Conclusion

The modulation of hyaluronidase enzyme and its substrate HA throughout the body is critical to maintain hyaluronan homeostasis as HA degradation is associated with pathogenesis of various health conditions. The literature survey carried out in this study found an increasing number of studies reported on HAase inhibitors derived from various biological sources and majority of the discoveries were from medicinal plants which have ethnobotanical claims for ailments associated with hyaluronan. Various class of natural products identified include alkaloids, flavonoids and terpenes have shown potent inhibitory activity against HAases in the *in vitro* studies. Similarly, a number of medicinal plant extracts and their fractions were found active against hyaluronidases and could serve as potential reservoirs for HAase inhibitors. These preliminary findings need further research to identify the active constituent(s) present in the extracts and establish their mechanism of action, safety profile and appropriate dosage of the active agents in animal and human studies. Hence, HAase inhibitors could be effective in controlling diseases involving uncontrolled degradation of HA and may serve as chondroprotective, anti-aging, antitumor, antimicrobial contraceptive agents, and anti-venom alternative.

Acknowledgements

The authors are thankful to the Department of Biological Sciences and NUMS Institute for Advanced Studies and Research (NIASR), National University of

Medical Sciences, Rawalpindi, Pakistan for supporting this study. We also apologize to the authors of many interesting studies that were omitted due to limitation.

Conflict of interest

The authors declare that they have no conflict of interest.

Author details

Muhammad Zeeshan Bhatti and Aman Karim*
Department of Biological Sciences, National University of Medical Sciences, Rawalpindi, Pakistan

*Address all correspondence to: aman.karim@numspak.edu.pk

References

[1] Laurent TC, Fraser JRE. Hyaluronan 1. The FASEB Journal. 1992;6(7):2397-404. DOI: 10.1096/fasebj.6.7.1563592

[2] Spicer AP, McDonald JA. Characterization and molecular evolution of a vertebrate hyaluronan synthase gene family. Journal of Biological Chemistry. 1998;273(4):1923-32. DOI: 10.1074/jbc.273.4.1923

[3] Lee JY, Spicer AP. Hyaluronan: a multifunctional, megaDalton, stealth molecule. Current Opinion in Cell Biology. 2000;12(5):581-6. DOI: 10.1016/s0955-0674(00)00135-6

[4] Tammi MI, Day AJ, Turley EA. Hyaluronan and homeostasis: a balancing act. Journal of Biological Chemistry. 2002;277(7):4581-4. DOI: 10.1074/jbc.R100037200

[5] Turley EA, Noble PW, Bourguignon LY. Signaling properties of hyaluronan receptors. Journal of Biological Chemistry. 2002;277(7):4589-92. DOI: 10.1074/jbc.R100038200

[6] Toole BP. Hyaluronan: from extracellular glue to pericellular cue. Nature Review Cancer. 2004;4(7):528-39. DOI: 10.1038/nrc1391

[7] Spicer AP, Tien JY. Hyaluronan and morphogenesis. Birth Defects Res C Embryo Today. 2004;72(1):89-108. DOI: 10.1002/bdrc.20006

[8] Adamia S, Maxwell CA, Pilarski LM. Hyaluronan and hyaluronan synthases: potential therapeutic targets in cancer. Curr Drug Targets Cardiovasc Haematol Disord. 2005;5(1):3-14. DOI: 10.2174/1568006053005056

[9] Meyer K. 11 Hyaluronidases. In: Boyer PD, editor. The Enzymes. The Enzymes. 5: Academic Press; 1971. p. 307-20. DOI: 10.1016/s1874-6047(08)60094-3

[10] Roden L, Campbell P, Fraser JR, Laurent TC, Pertoft H, Thompson JN. Enzymic pathways of hyaluronan catabolism. Ciba Found Symp. 1989;143:60-76; discussion -86, 281-5. DOI: 10.1002/9780470513774.ch5

[11] Frost GI, Csóka T, Stern R. The Hyaluronidases: A Chemical, Biological and Clinical Overview. Trends in Glycoscience and Glycotechnology. 1996;8(44):419-34. DOI: 10.4052/tigg.8.419

[12] Noble PW. Hyaluronan and its catabolic products in tissue injury and repair. Matrix Biology. 2002;21(1):25-9. DOI: 10.1016/s0945-053x(01)00184-6

[13] Menzel EJ, Farr C. Hyaluronidase and its substrate hyaluronan: biochemistry, biological activities and therapeutic uses. Cancer Letters. 1998;131(1):3-11. DOI: 10.1016/s0304-3835(98)00195-5

[14] Heldin P. Importance of hyaluronan biosynthesis and degradation in cell differentiation and tumor formation. Brazilian Journal of Medical and Biological Research. 2003;36(8):967-73. DOI: 10.1590/s0100-879x2003000800002

[15] Termeer C, Sleeman JP, Simon JC. Hyaluronan–magic glue for the regulation of the immune response? Trends in Immunology. 2003;24(3):112-4. DOI: 10.1016/s1471-4906(03)00029-2

[16] George J, Stern R. Serum hyaluronan and hyaluronidase: very early markers of toxic liver injury. Clinica Chimica Acta. 2004;348(1-2):189-97. DOI: 10.1016/j.cccn.2004.05.018

[17] Day AJ, de la Motte CA. Hyaluronan cross-linking: a protective mechanism in inflammation? Trends in Immunology. 2005;26(12):637-43. DOI: 10.1016/j.it.2005.09.009

[18] Jiang D, Liang J, Fan J, Yu S, Chen S, Luo Y, et al. Regulation of lung injury and repair by Toll-like receptors and hyaluronan. Nature Medicine. 2005; 11(11):1173-9. DOI: 10.1038/nm1315

[19] Chen WY, Abatangelo G. Functions of hyaluronan in wound repair. Wound Repair and Regeneration. 1999;7(2):79-89. DOI: 10.1046/j.1524-475x.1999.00079.x

[20] West DC, Hampson IN, Arnold F, Kumar S. Angiogenesis induced by degradation products of hyaluronic acid. Science. 1985;228(4705):1324-6. DOI: 10.1126/science.2408340

[21] McDonald JA, Camenisch TD. Hyaluronan: genetic insights into the complex biology of a simple polysaccharide. Glycoconjugate Journal. 2002;19(4-5):331-9. DOI: 10.1023/A:1025369004783

[22] Girish KS, Kemparaju K. The magic glue hyaluronan and its eraser hyaluronidase: a biological overview. Life Science. 2007;80(21):1921-43. DOI: 10.1016/j.lfs.2007.02.037

[23] Mio K, Stern R. Inhibitors of the hyaluronidases. Matrix Biology. 2002;21(1):31-7. DOI: 10.1016/s0945-053x(01)00185-8

[24] Girish KS, Kemparaju K, Nagaraju S, Vishwanath BS. Hyaluronidase inhibitors: a biological and therapeutic perspective. Current Medicinal Chemistry. 2009;16(18):2261-88. DOI: 10.2174/092986709788453078

[25] Khanum SA, Murari SK, Vishwanth BS, Shashikanth S. Synthesis of benzoyl phenyl benzoates as effective inhibitors for phospholipase A2 and hyaluronidase enzymes. Bioorganic Medicinal Chemistry Letters. 2005;15(18):4100-4. DOI: 10.1016/j.bmcl.2005.06.012

[26] Salmen S, Hoechstetter J, Kasbauer C, Paper DH, Bernhardt G, Buschauer A. Sulphated oligosaccharides as inhibitors of hyaluronidases from bovine testis, bee venom and *Streptococcus agalactiae*. Planta Medica. 2005;71(8):727-32. DOI: 10.1055/s-2005-871255

[27] Girish KS, Kemparaju K. A low molecular weight isoform of hyaluronidase: purification from Indian cobra (*Naja naja*) venom and partial characterization. Biochemistry (Mosc). 2005;70(6):708-12. DOI: 10.1007/s10541-005-0172-6

[28] Girish KS, Kemparaju K. Inhibition of *Naja naja* venom hyaluronidase by plant-derived bioactive components and polysaccharides. Biochemistry (Mosc). 2005;70(8):948-52. DOI: 10.1007/s10541-005-0207-z

[29] Isoyama T, Thwaites D, Selzer MG, Carey RI, Barbucci R, Lokeshwar VB. Differential selectivity of hyaluronidase inhibitors toward acidic and basic hyaluronidases. Glycobiology. 2006;16(1):11-21. DOI: 10.1093/glycob/cwj036

[30] Machiah DK, Girish KS, Gowda TV. A glycoprotein from a folk medicinal plant, *Withania somnifera*, inhibits hyaluronidase activity of snake venoms. Comparative Biochemistry and Physiology. Toxicology & Pharmacology. 2006;143(2):158-61. DOI: 10.1016/j.cbpc.2006.01.006

[31] Kaul, A., Short, W. D., Wang, X., & Keswani, S. G. Hyaluronidases in Human Diseases. International Journal of Molecular Sciences. 2021;22(6), 3204.

[32] Duran-Reynals, F. (1929). The effect of extracts of certain organs from normal and immunized animals on the infecting power of vaccine virus. The Journal of Experimental Medicine. 50(3), 327-340.

[33] Cramer, J. A., Bailey, L. C., Bailey, C. A., & Miller, R. T. Kinetic and

mechanistic studies with bovine testicular hyaluronidase. Biochimica et Biophysica Acta (BBA)-General Subjects. 1994;1200(3), 315-321.

[34] Hotez, P. J., Narasimhan, S., Haggerty, J., Milstone, L., Bhopale, V., Schad, G. A., & Richards, F. F. Hyaluronidase from infective Ancylostoma hookworm larvae and its possible function as a virulence factor in tissue invasion and in cutaneous larva migrans. Infection and Immunity. (1992); 60(3),1018-1023.

[35] Makris, G., Wright, J. D., Ingham, E., & Holland, K. T.. The hyaluronate lyase of Staphylococcus aureus–a virulence factor?. Microbiology. 2004;150(6), 2005-2013.

[36] Csóka, A. B., Scherer, S. W., & Stern, R.. Expression analysis of six paralogous human hyaluronidase genes clustered on chromosomes 3p21 and 7q31. Genomics. 999;60(3), 356-361.

[37] Csoka, A. B., Frost, G. I., & Stern, R. The six hyaluronidase-like genes in the human and mouse genomes. Matrix Biology. 2001;20(8), 499-508.

[38] Mio, K., & Stern, R. Inhibitors of the hyaluronidases. Matrix Biology. 2002;21(1), 31-37.

[39] Mio, K., Carrette, O., Maibach, H. I., & Stern, R. Evidence that the serum inhibitor of hyaluronidase may be a member of the inter-α-inhibitor family. Journal of Biological Chemistry. 2000;275(42), 32413-32421.

[40] Newman, D. J. Natural products as leads to potential drugs: an old process or the new hope for drug discovery?. Journal of Medicinal Chemistry. 2008;51(9), 2589-2599.

[41] Harvey, A. L., Edrada-Ebel, R., & Quinn, R. J. The re-emergence of natural products for drug discovery in the genomics era. Nature Reviews Drug Discovery. (2015);14(2), 111-129.

[42] Lahlou, M. The success of natural products in drug discovery. Pharmacology & Pharmacy. 2013;4(3), Aticle ID:33502. DOI:10.4236/pp.2013.43A003

[43] Ganesan, A. (2008). The impact of natural products upon modern drug discovery. Current opinion in chemical biology, 12(3), 306-317.

[44] Ortholand, J. Y., & Ganesan, A. (2004). Natural products and combinatorial chemistry: back to the future. Current opinion in chemical biology, 8(3), 271-280.

[45] Morikawa T, Okugawa S, Manse Y, Muraoka O, Yoshikawa M, Ninomiya K. Quantitative determination of principal aporphine and benzylisoquinoline alkaloids due to blooming state in lotus flower (flower buds of *Nelumbo nucifera*) and their hyaluronidase inhibitory activity. Natural Product Communications. 2019;14(6). DOI: 10.1177/1934578x19857834

[46] Girish KS, Kemparaju K. Inhibition of *Naja naja* venom hyaluronidase: role in the management of poisonous bite. Life Science. 2006;78(13):1433-40. DOI: 10.1016/j.lfs.2005.07.015

[47] Olgen S, Kaessler A, Kilic-Kurt Z, Jose J. Investigation of aminomethyl indole derivatives as hyaluronidase inhibitors. Zeitschrift für Naturforschung C Journal of Biosciences. 2010;65(7-8): 445-50. DOI: 10.1515/znc-2010-7-805

[48] Havsteen B. Flavonoids, a class of natural products of high pharmacological potency. Biochemical Pharmacology. 1983;32(7):1141-8. DOI: 10.1016/0006-2952(83)90262-9

[49] Hollman PCH, Arts ICW. Flavonols, flavones and flavanols-nature, occurrence and dietary burden. Journal of the Science of Food and Agriculture. 2000;80(7):1081-93. DOI: 10.1002/(sici)1097-0010(20000515)80:7<1081::Aid-jsfa566>3.0.Co;2-g

[50] Middleton E, Jr., Kandaswami C, Theoharides TC. The effects of plant flavonoids on mammalian cells: implications for inflammation, heart disease, and cancer. Pharmacological reviews. 2000;52(4):673-751. DOI:

[51] Rodney G, Swanson AL, Wheeler LM, Smith GN, Worrel CS. The effect of a series of flavonoids on hyaluronidase and some other related enzymes. Journal of Biological Chemistry. 1950;183:739-47. DOI:

[52] Hertel W, Peschel G, Ozegowski JH, Muller PJ. Inhibitory effects of triterpenes and flavonoids on the enzymatic activity of hyaluronic acid-splitting enzymes. Archiv der Pharmazie (Weinheim). 2006;339(6):313-8. DOI: 10.1002/ardp.200500216

[53] Kuppusamy UR, Khoo HE, Das NP. Structure-activity studies of flavonoids as inhibitors of hyaluronidase. Biochemical Pharmacology. 1990;40(2):397-401. DOI: 10.1016/0006-2952(90)90709-t

[54] Kuppusamy UR, Das NP. Protective effects of tannic acid and related natural compounds on *Crotalus adamenteus* subcutaneous poisoning in mice. Pharmacology and Toxicology. 1993;72(4-5):290-5. DOI: 10.1111/j.1600-0773.1993.tb01652.xK

[55] im MK, Kim Y, Chung S. Identification and *in vitro* biological activities of flavonols in garlic leaf and shoot: inhibition of soybean lipoxygenase and hyaluronidase activities and scavenging of free radicals. Journal of the Science of Food and Agriculture. 2005;85(4):633-40. DOI: 10.1002/jsfa.1899

[56] Pandey KB, Rizvi SI. Plant polyphenols as dietary antioxidants in human health and disease. Oxidative Medicine and Cellular Longevity. 2009;2(5):270-8. DOI: 10.4161/oxim.2.5.9498

[57] Murata T, Miyase T, Yoshizaki F. Cyclic spermidine alkaloids and flavone glycosides from *Meehania fargesii*. Chemical Pharmacetical Bulletin (Tokyo). 2010;58(5):696-702. DOI: 10.1248/cpb.58.696

[58] Tatemoto H, Tokeshi I, Nakamura S, Muto N, Nakada T. Inhibition of boar sperm hyaluronidase activity by tannic acid reduces polyspermy during in vitro fertilization of porcine oocytes. Zygote. 2006;14(4):275-85. DOI: 10.1017/S0967199406003819

[59] Tokeshi I, Yoshimoto T, Muto N, Nakamura S, Ashizawa K, Nakada T, et al. Antihyaluronidase action of ellagic acid effectively prevents polyspermy as a result of suppression of the acrosome reaction induced by sperm-zona interaction during in vitro fertilization of porcine oocytes. Journal of Reproductive Development. 2007;53(4):755-64. DOI: 10.1262/jrd.18173

[60] Kiyama R. Estrogenic terpenes and terpenoids: Pathways, functions and applications. European Journal Pharmacology. 2017;815:405-15. DOI: 10.1016/j.ejphar.2017.09.049

[61] Lim SH, Ha TY, Ahn J, Kim S. Estrogenic activities of *Psoralea corylifolia* L. seed extracts and main constituents. Phytomedicine. 2011;18(5):425-30. DOI: 10.1016/j.phymed.2011.02.002

[62] Abdullah NH, Thomas NF, Sivasothy Y, Lee VS, Liew SY, Noorbatcha IA, et al. Hyaluronidase inhibitory activity of pentacylic triterpenoids from *Prismatomeris tetrandra* (Roxb.) K. Schum: Isolation, synthesis and QSAR study. International Journal of Molecular Sciences. 2016;17(2):143. DOI: 10.3390/ijms17020143

[63] Botzki A, Rigden DJ, Braun S, Nukui M, Salmen S, Hoechstetter J, et al. L-Ascorbic acid 6-hexadecanoate, a potent hyaluronidase inhibitor. X-ray structure

and molecular modeling of enzyme-inhibitor complexes. Journal of Biological Chemistry. 2004;279(44):45990-7. DOI: 10.1074/jbc.M406146200

[64] Rodig H, Ozegowski JH, Peschel G, Müller PJ. Complementary characterization of a hyaluronic acid splitting enzyme from *streptococcus agalactiae*. Zentralblatt für Bakteriologie. 2000;289(8):835-43. DOI: 10.1016/s0934-8840(00)80011-0

[65] Panda S, Jafri M, Kar A, Meheta BK. Thyroid inhibitory, antiperoxidative and hypoglycemic effects of stigmasterol isolated from *Butea monosperma*. Fitoterapia. 2009;80(2):123-6. DOI: 10.1016/j.fitote.2008.12.002

[66] Papakonstantinou E, Klagas I, Karakiulakis G, Hostettler K, S'Ng C T, Kotoula V, et al. Steroids and beta2-agonists regulate hyaluronan metabolism in asthmatic airway smooth muscle cells. America Journal of Respiratory Cell and Molecular Biology. 2012;47(6):759-67. DOI: 10.1165/rcmb.2012-0101OC

[67] Patil S, Bhadane B, Shirsath L, Patil R, Chaudhari B. Steroidal fraction of *Carissa carandas* L. inhibits microbial hyaluronidase activity by mixed inhibition mechanism. Preparative Biochemistry and Biotechnology. 2019;49(3):298-306. DOI: 10.1080/10826068.2018.1541811

[68] Lengers I, Herrmann F, Le Borgne M, Jose J. Improved surface display of human Hyal1 and identification of testosterone propionate and chicoric acid as new inhibitors. Pharmaceuticals (Basel). 2020;13(4). DOI:10.3390/ph13040054.

[69] Facino RM, Carini M, Stefani R, Aldini G, Saibene L. Anti-elastase and anti-hyaluronidase activities of saponins and sapogenins from *Hedera helix*, *Aesculus hippocastanum*, and *Ruscus aculeatus*: factors contributing to their efficacy in the treatment of venous insufficiency. Archiv der Pharmazie (Weinheim). 1995;328(10):720-4. DOI: 10.1002/ardp.19953281006

[70] Tekulu GH, Desta A, Hiben MG, Araya EM. Anti-nociceptive and anti-inflammatory activity of *Hygrophila schulli* leaves. Journal of Inflammatory Research. 2020;13:497-505. DOI: 10.2147/JIR.S269717

[71] Lee KK, Kim JH, Cho JJ, Choi JD. Inhibitory Effects of 150 plant extracts on elastase activity, and their anti-inflammatory effects. International Journal of Cosmetic Sciences. 1999;21(2):71-82. DOI: 10.1046/j.1467-2494.1999.181638.x

[72] Wang C, Lee W. Separation, characteristics, and biological activities of phenolics in areca fruit. Journal of Agricultural and Food Chemistry. 1996;44(8):2014-9. DOI: 10.1021/jf950611o

[73] Tomohara K, Ito T, Onikata S, Kato A, Adachi I. Discovery of hyaluronidase inhibitors from natural products and their mechanistic characterization under DMSO-perturbed assay conditions. Bioorganic and Medicinal Chemistry Letters. 2017;27(7):1620-3. DOI: 10.1016/j.bmcl.2017.01.083

[74] Jeong SJ, Ahn NH, Kim YC, Inagaki M, Miyamoto T, Higuchi R. Norlignans with Hyaluronidase Inhibitory Activity from A*nemarrhena asphodeloides*. Planta Medica. 1999;65(4):367-8. DOI: 10.1055/s-2006-960789

[75] Marquina MA, Corao GM, Araujo L, Buitrago D, Sosa M. Hyaluronidase inhibitory activity from the polyphenols in the fruit of blackberry (*Rubus fruticosus* B.). Fitoterapia. 2002;73(7-8):727-9. DOI: 10.1016/s0367-326x(02)00222-8

[76] Liyanaarachchi GD, Samarasekera JKRR, Mahanama KRR, Hemalal KDP. Tyrosinase, elastase, hyaluronidase, inhibitory and antioxidant

activity of Sri Lankan medicinal plants for novel cosmeceuticals. Industrial Crops and Products. 2018;111:597-605. DOI: 10.1016/j.indcrop.2017.11.019

[77] Selenge E, Odontuya G, Murata T, Sasaki K, Kobayashi K, Batkhuu J, et al. Phytochemical constituents of Mongolian traditional medicinal plants, *Chamaerhodos erecta* and *C. altaica*, and its constituents prevents the extracellular matrix degradation factors. Journal of Natural Medicines. 2013;67(4):867-75. DOI: 10.1007/s11418-013-0748-1

[78] Ratnasooriya WD, Abeysekera WPKM, Ratnasooriya CTD. *In vitro* anti-hyaluronidase activity of Sri Lankan low grown orthodox orange pekoe grade black tea (*Camellia sinensis* L.). Asian Pacific Journal of Tropical Biomedicine. 2014;4(12):959-63. DOI: 10.12980/apjtb.4.2014apjtb-2014-0462

[79] Han SS, Hur SJ, Lee SK. A comparison of antioxidative and anti-inflammatory activities of sword beans and soybeans fermented with *Bacillus subtilis*. Food Funct. 2015;6(8):2736-48. DOI: 10.1039/c5fo00290g

[80] Furusawa M, Narita Y, Iwai K, Fukunaga T, Nakagiri O. Inhibitory effect of a hot water extract of coffee "silverskin" on hyaluronidase. Bioscience, Biotechnology and Biochemistry. 2011;75(6):1205-7. DOI: 10.1271/bbb.110106

[81] Hwang SG, Yang A, Kim SJ, Kim MK, Kim SS, Oh HJ, et al. Screening of hyaluronidase inhibitor in Korean medicinal plants. Journal of Life Science. 2014;24(5):498-504. DOI: 10.5352/jls.2014.24.5.498

[82] Selenge E, Murata T, Tanaka S, Sasaki K, Batkhuu J, Yoshizaki F. Monoterpene glycosides, phenylpropanoids, and acacetin glycosides from *Dracocephalum foetidum*. Phytochemistry. 2014;101:91-100. DOI: 10.1016/j.phytochem.2014.02.007

[83] Zaluski D, Janeczko Z. Variation in phytochemicals and bioactivity of the fruits of Eleutherococcus species cultivated in Poland. Natural Product Research. 2015;29(23):2207-11. DOI: 10.1080/14786419.2014.1002091

[84] Panossian A, Wikman G, Wagner H. Plant adaptogens III. Earlier and more recent aspects and concepts on their mode of action. Phytomedicine. 1999;6(4):287-300. DOI: 10.1016/s0944-7113(99)80023-3

[85] Jin L, Wu F, Li X, Li H, Du C, Jiang Q, et al. Anti-depressant effects of aqueous extract from *Acanthopanax senticosus* in mice. Phytotherapy Research. 2013;27(12):1829-33. DOI: 10.1002/ptr.4938

[86] Guo S, Liu Y, Lin Z, Tai S, Yin S, Liu G. Effects of eleutheroside B and eleutheroside E on activity of cytochrome P450 in rat liver microsomes. BMC Complementary and Alternative Medicine. 2014;14:1. DOI: 10.1186/1472-6882-14-1

[87] Murata T, Miyase T, Yoshizaki F. Hyaluronidase inhibitors from *Keiskea japonica*. Chemistry and Pharmaceutica; Bulletin (Tokyo). 2012;60(1):121-8. DOI: 10.1248/cpb.60.121

[88] Murata T, Watahiki M, Tanaka Y, Miyase T, Yoshizaki F. Hyaluronidase inhibitors from Takuran, *Lycopus lucidus*. Chemistry and Pharmaceutical Bulletin (Tokyo). 2010;58(3):394-7. DOI: 10.1248/cpb.58.394

[89] Piwowarski JP, Kiss AK, Kozlowska-Wojciechowska M. Anti-hyaluronidase and anti-elastase activity screening of tannin-rich plant materials used in traditional Polish medicine for external treatment of diseases with inflammatory background. Journal of Ethnopharmacology. 2011;137(1):937-41. DOI: 10.1016/j.jep.2011.05.039

[90] Srivastav A, Chandra A, Singh M, Jamal F, Rastogi P, Rajendran SM,

Bansode FW, Lakshmi V. Inhibition of hyaluronidase activity of human and rat spermatozoa in vitro and antispermatogenic activity in rats in vivo by *Terminalia chebula*, a flavonoid rich plant. Reproductive Toxicology. 2010;29:214-24.

[91] Michel P, Owczarek A, Matczak M, Kosno M, Szymański P, Mikiciuk-Olasik E, Kilanowicz A, Wesołowski W, Olszewska MA. Metabolite profiling of eastern teaberry (*Gaultheria procumbens* L.) lipophilic leaf extracts with hyaluronidase and lipoxygenase inhibitory activity. Molecules. 201;22:412. DOI: 10.3390/molecules22030412

[92] Citalingam K, Zareen S, Shaari K, Ahmad S. Effects of *Payena dasyphylla* (Miq.) on hyaluronidase enzyme activity and metalloproteinases protein expressions in interleukin-1beta stimulated human chondrocytes cells. BMC Complementary and Alternative Medicine. 2013;13:213. DOI: 10.1186/1472-6882-13-213

[93] Sumantran VN, Kulkarni A, Chandwaskar R, Harsulkar A, Patwardhan B, Chopra A, et al. Chondroprotective potential of fruit extracts of *Phyllanthus emblica* in osteoarthritis. Evidence Based Complementary and Alternaternative Medicine. 2008;5(3):329-35. DOI: 10.1093/ecam/nem030

[94] Bralley E, Greenspan P, Hargrove JL, Hartle DK. Inhibition of Hyaluronidase activity by *Vitis rotundifolia*. (Muscadine) berry seeds and skins. Pharmaceutical Biology. 2008;45(9):667-73. DOI: 10.1080/13880200701545018

[95] Bose B, Choudhury H, Tandon P, Kumaria S. Studies on secondary metabolite profiling, anti-inflammatory potential, *in vitro* photoprotective and skin-aging related enzyme inhibitory activities of *Malaxis acuminata*, a threatened orchid of nutraceutical importance. Journal of Photochemistry and Photobiology B. 2017;173:686-95. DOI: 10.1016/j.jphotobiol.2017.07.010

[96] Girish KS, Mohanakumari HP, Nagaraju S, Vishwanath BS, Kemparaju K. Hyaluronidase and protease activities from Indian snake venoms: neutralization by *Mimosa pudica* root extract. Fitoterapia. 2004;75(3-4):378-80. DOI: 10.1016/j.fitote.2004.01.006

[97] Granica S, Czerwinska ME, Piwowarski JP, Ziaja M, Kiss AK. Chemical composition, antioxidative and anti-inflammatory activity of extracts prepared from aerial parts of *Oenothera biennis* L. and *Oenothera paradoxa* Hudziok obtained after seeds cultivation. Journal of Agriculture and Food Chemistry. 2013;61(4):801-10. DOI: 10.1021/jf304002h

[98] Bahta T, Karim A, Periasamy G, Gebremedhin G, Ur-Rehman N, Bitew H, et al. Analgesic, anti-inflammatory and in-vitro hyaluronidase inhibitory properties of the leaf extract and solvent fractions of *Otostegia fruticosa* (Forssk.) Schweinf. ex Penzig. Iranian Journal of Pharmaceutical Research. 2020;19(1):218-30. DOI: 10.22037/ijpr.2019.14657.12569

[99] Shibata T, Fujimoto K, Nagayama K, Yamaguchi K, Nakamura T. Inhibitory activity of brown algal phlorotannins against hyaluronidase. International journal of Food Science and Technology. 2002;37:703-709. DOI: 10.1046/j.1365-2621.2002.00603.x

[100] Fayad S, Nehme R, Tannoury M, Lesellier E, Pichon C, Morin P. Macroalga *Padina pavonica* water extracts obtained by pressurized liquid extraction and microwave-assisted extraction inhibit hyaluronidase activity as shown by capillary electrophoresis. Journal of Chromatography A. 2017;1497:19-27. DOI: 10.1016/j.chroma.2017.03.033